SpringerBriefs in Mathematics

SpringerBriefs present concise summaries of cutting-edge research and practical applications across a wide spectrum of fields. Featuring compact volumes of 50 to 125 pages, the series covers a range of content from professional to academic. Briefs are characterized by fast, global electronic dissemination, standard publishing contracts, standardized manuscript preparation and formatting guidelines, and expedited production schedules.

Typical topics might include:

A timely report of state-of-the art techniques A bridge between new research results, as published in journal articles, and a contextual literature review A snapshot of a hot or emerging topic An in-depth case study A presentation of core concepts that students must understand in order to make independent contributions

SpringerBriefs in Mathematics showcases expositions in all areas of mathematics and applied mathematics. Manuscripts presenting new results or a single new result in a classical field, new field, or an emerging topic, applications, or bridges between new results and already published works, are encouraged. The series is intended for mathematicians and applied mathematicians. All works are peer-reviewed to meet the highest standards of scientific literature.

Titles from this series are indexed by Scopus, Web of Science, Mathematical Reviews, and zbMATH.

More information about this series at http://www.springer.com/series/10030

Toshio Mikami

Stochastic Optimal Transportation

Stochastic Control with Fixed Marginals

Springer

Toshio Mikami
Department of Mathematics
Tsuda University
Tokyo, Japan

ISSN 2191-8198 ISSN 2191-8201 (electronic)
SpringerBriefs in Mathematics
ISBN 978-981-16-1753-9 ISBN 978-981-16-1754-6 (eBook)
https://doi.org/10.1007/978-981-16-1754-6

This Springer imprint is published by the registered company Springer Nature Singapore Pte Ltd.
The registered company address is: 152 Beach Road, #21-01/04 Gateway East, Singapore 189721, Singapore

Preface

The construction of a stochastic process from given marginal distributions is an important part of the so-called marginal problem. A Markovian Bernstein process on [0, 1] constructed from two endpoint marginals at times $t = 0, 1$ and from the Bernstein transition density solves Schrödinger's functional equation. It is the so-called h-path process, provided the Markov process solves a stochastic differential equation. The Markov diffusion process and its construction from a solution of the Fokker–Planck equation are called the Nelson process and Nelson's problem, respectively. (In E. Nelson's original problem, he considered the Fokker–Planck equation which is satisfied by the square of the absolute value of a solution to Schrödinger's equation.) In the second part of our dissertation under the supervision of Professor Wendell H. Fleming, Brown University, we gave an approach to Nelson's problem via the continuum limit of a class of stochastic controls with given two endpoint marginals. It is closely related to the optimal transportation problem, in that we find optimal dynamics, in the minimization problem, among those who have the same partial information on marginal distributions. Nearly 15 years later, we gave a probabilistic proof to Monge's problem with a quadratic cost by the zero-noise limit of h-path processes, which encouraged us to consider optimal transportation for semimartingales which we call the **stochastic optimal transportation**. It can be considered a class of marginal problems.

In Chap. 1, we introduce Monge's problem and Schrödinger's to compare them so that one can see the relation between the nonstochastic and stochastic optimal transportations.

The Duality Theorems for stochastic optimal transportation problems are useful for considering marginal problems, including Nelson's. Indeed, we used them to construct a semimartingale from the Fokker–Planck equation under a general integrability condition. The construction of a semimartingale from the Fokker–Planck equation is also called the superposition principle and has been remarkably developed in the last several years.

In Chap. 2, we give two classes of stochastic optimal transportation problems. As an application, by the superposition principle, we give our recent progress on the Duality Theorems for stochastic optimal transportation problems with a convex cost function. We also give a sufficient condition for the finiteness of the minimum in stochastic optimal transportation problem and discuss the relation between the nonstochastic and stochastic optimal transportations by the zero-noise limit.

In Chap. 3, we consider the finiteness, the semiconcavity, and the continuity of the minimum in Schrödinger's problem. We also consider the regularity of the solution to Schrödinger's functional equation, marginal problems for stochastic processes, and stochastic optimal transportation with a nonconvex cost function in the one-dimensional case.

The generalizations of the results in Chap. 3 are our future projects.

Besides the page limit, since it is beyond our ability to introduce all topics in this rapidly developing field, we ask readers to find papers on missing topics such as stochastic mechanics, large deviations, entropic functional inequalities, martingale optimal transports, etc.

We would like to thank anonymous referees for constructive suggestions and for informing us about missing references, and thank Mr. Masayuki Nakamura, Springer for his support. We would also like to acknowledge the financial support by JSPS KAKENHI Grant Numbers JP16H03948 and 19K03548.

I would like to thank Professors Wendell H. Fleming, Hitoshi Ishii, and Masayoshi Takeda for their academic influences and constant encouragement since I was a graduate student.

Lastly, I would like to thank my wife Mari for her unconditional support in my lifetime.

Tokyo, Japan Toshio Mikami
February 2021

Contents

Notation

OT	Optimal transportation problem
SOT	Stochastic optimal transportation problem
FBSDE	Forward–backward stochastic differential equation
HJB	Hamilton–Jacobi–Bellman
$\mathbf{B}(S)$	Set of all Borel measurable subsets of a topological space S
$\mathcal{P}(S)$	Set of all Borel probability measures on a topological space S
$\mathcal{P}_{ac}(\mathbb{R}^d)$	$\{P \in \mathcal{P}(\mathbb{R}^d) \| P(dx) \ll dx\}$
$\mathcal{P}_r(\mathbb{R}^d)$	$\{P \in \mathcal{P}(\mathbb{R}^d) \| \int_{\mathbb{R}^d} \|x\|^r P(dx) < \infty\}$ for $r \geq 1$
$\mathcal{P}_{r,ac}(\mathbb{R}^d)$	$\mathcal{P}_{ac}(\mathbb{R}^d) \cap \mathcal{P}_r(\mathbb{R}^d)$ for $r \geq 1$
$\mathcal{A}(P_0, P_1)$	$\{\mu \in \mathcal{P}(\mathbb{R}^{2d}) \| \mu(dx \times \mathbb{R}^d) = P_0(dx), \mu(\mathbb{R}^d \times dx) = P_1(dx)\}$
$AC(S)$	Set of all absolutely continuous functions on S
$USC(S)$	Set of all upper semicontinuous functions on S
$LSC(S)$	Set of all lower semicontinuous functions on S
$UC_b(S)$	Set of all uniformly continuous bounded functions on S
$C^{i,j}(S)$	Set of all functions on S, which are ith and jth continuously differentiable in the first and the second variables, respectively
$C_b^{i,j}(S)$	Set of all functions on S, which have bounded continuous derivatives up to the ith and jth orders in the first and the second variables, respectively
P^X	Probability distribution of a random variable X
$\dot{x}(t)$	$dx(t)/dt$
$\langle x, y \rangle$	$\sum_{i=1}^d x_i y_i$ for $x = (x_i)_{i=1}^d, y = (y_i)_{i=1}^d \in \mathbb{R}^d$
$\langle A, B \rangle$	$\sum_{i,j=1}^d a_{ij} b_{ij}$ for $A = (a_{ij})_{i,j=1}^d, B = (b_{ij})_{i,j=1}^d \in M(d, \mathbb{R})$
D_x	$(\partial/\partial x_i)_{i=1}^d$
D_x^2	$(\partial^2/\partial x_i \partial x_j)_{i,j=1}^d$
$\|\|f\|\|_\infty$	$\sup_{x \in S} \|f(x)\|, \quad f \in C(S)$
B_r	$\{x \in \mathbb{R}^d : \|x\| \leq r\}$

$H(P\|Q)$	$= \int_S \{\log(P(dx)/Q(dx))\} P(dx)$ if $P \ll Q$; $= \infty$, otherwise.
$\mathscr{S}(P)$	$= \int_{\mathbb{R}^d} \{\log p(x)\} p(x)dx$ if $P(dx) = p(x)dx$; $= \infty$, otherwise.
$\|\|x\|\|_{L^2(P)}$	$\sqrt{\int_{\mathbb{R}^d} \|x\|^2 P(dx)}$, $P \in \mathscr{P}_2(\mathbb{R}^d)$
(A.0.0)	(i) $\sigma_{ij} \in C_b([0,1] \times \mathbb{R}^d)$, $i, j = 1, \cdots, d$. (ii) σ is nondegenerate.
(A.0)	$\sigma_{ij} \in C_b^1([0,1] \times \mathbb{R}^d)$, $i, j = 1, \cdots, d$.
(A.1)	(i) $L \in C([0,1] \times \mathbb{R}^d \times \mathbb{R}^d; [0,\infty))$. (ii) For $(t,x) \in [0,1] \times \mathbb{R}^d$, $u \mapsto L(t,x;u)$ is convex.
(A.2)	$\lim_{\|u\| \to \infty} \inf\{L(t,x;u) \| (t,x) \in [0,1] \times \mathbb{R}^d\}/\|u\| = \infty$.
(A.3)	(i) $\partial L(t,x;u)/\partial t$ and $D_x L(t,x;u)$ are bounded on $[0,1] \times \mathbb{R}^d \times B_R$ for all $R > 0$, where $B_R := \{x \in \mathbb{R}^d \| \|x\| \le R\}$. (ii) C_L is finite, where $C_L := \sup\{L(t,x;u)/(1 + L(t,y;u)) \| 0 \le t \le 1, x, y, u \in \mathbb{R}^d\}$.
(A.4)	(i) "σ is an identity" or "σ is uniformly nondegenerate, $\sigma_{ij} \in C_b^{1,2}([0,1] \times \mathbb{R}^d)$, $i, j = 1, \cdots, d$ and there exist functions L_1 and L_2 so that $L = L_1(t,x) + L_2(t,u)$". (ii) $L(t,x;u) \in C^1([0,1] \times \mathbb{R}^d \times \mathbb{R}^d; [0,\infty))$ and is strictly convex in u. (iii) $L \in C_b^{1,2,0}([0,1] \times \mathbb{R}^d \times B_R)$ for any $R > 0$.
(A.4)'	σ is uniformly nondegenerate, $\sigma_{ij} \in C_b^{1,2}([0,1] \times \mathbb{R}^d)$, $i, j = 1, \cdots, d$. $\xi \in C_b^{1,2}([0,1] \times \mathbb{R}^d; \mathbb{R}^d)$, $c \in C_b^{1,2}([0,1] \times \mathbb{R}^d)$, and for $(t,x,u) \in [0,1] \times \mathbb{R}^d \times \mathbb{R}^d$, $L(t,x;u) = \frac{1}{2}\langle a(t,x)^{-1}(u - \xi(t,x)), u - \xi(t,x)\rangle + c(t,x)$.
(A.5)	$L(t,x;\cdot) \in C^2(\mathbb{R}^d)$ for $(t,x) \in [0,1] \times \mathbb{R}^d$. $D_u^2 L(t,x;u)$ is bounded and uniformly nondegenerate on $[0,1] \times \mathbb{R}^d \times \mathbb{R}^d$.
(A.6)	$\sigma(t,x) = (\sigma^{ij}(t,x))_{i,j=1}^d$, $(t,x) \in [0,1] \times \mathbb{R}^d$, is a $d \times d$-matrix. $a := \sigma\sigma^t$ is uniformly positive definite, bounded, once continuously differentiable and uniformly Hölder continuous. $D_x a$ is bounded and the first derivatives of a are uniformly Hölder continuous in x uniformly in $t \in [0,1]$.
(A.7)	$\xi \in C_b([0,1] \times \mathbb{R}^d; \mathbb{R}^d)$ and is uniformly Hölder continuous in x uniformly in $t \in [0,1]$.
(A.8)	S is a compact metric space.
(A.8)'	S is a σ-compact metric space.
(A.8)"	S is a complete σ-compact metric space.
(A.9)	$q \in C(S \times S; (0,\infty))$.
(A.9.r)'	There exists $C_r > 0$ for which $x \mapsto C_r\|x\|^2 + \log q(x,y)$ and $y \mapsto C_r\|y\|^2 + \log q(x,y)$ are convex on B_r for any y and $x \in B_r$, respectively.
(A.10)	$L : \mathbb{R}^d \longrightarrow [0,\infty)$ is convex and $\liminf_{\|u\| \to \infty} L(u)/\|u\|^2 > 0$.
$V(P_0, P_1)$	$\inf\{E[\int_0^1 L(t, X(t); \beta_X(t,X))dt] \| X \in \mathscr{A}, P^{X(t)} = P_t, t = 0, 1\}$
$\mathrm{v}(P_0, P_1)$	$\inf\{\int_{[0,1]\times\mathbb{R}} L(t,x;b(t,x))dt Q_t(dx) \| b \in \mathbf{A}(\{Q_t\}_{0\le t\le 1}), Q_t = P_t, t = 0, 1\}$
$\tilde{\mathrm{v}}(P_0, P_1)$	$\inf\{\int_{[0,1]\times\mathbb{R}^d\times\mathbb{R}^d} L(t,x;u)\nu(dtdxdu) \| \nu \in \tilde{\mathscr{A}}, \nu_{1,t} = P_t, t = 0, 1\}$

$\mathbf{V}(\{P_t\}_{0\le t\le 1})$	$\inf\{E[\int_0^1 L(t, X(t); \beta_X(t, X))dt] \mid X \in \mathscr{A}, P^{X(t)} = P_t, 0 \le t \le 1\}$
$\mathbf{v}(\{P_t\}_{0\le t\le 1})$	$\inf\{\int_{[0,1]\times\mathbb{R}} L(t, x; b(t, x))dt P_t(dx) \mid b \in \mathbf{A}(\{P_t\}_{0\le t\le 1})\}$
$\tilde{\mathbf{v}}(\{P_t\}_{0\le t\le 1})$	$\inf\{\int_{[0,1]\times\mathbb{R}^d\times\mathbb{R}^d} L(t, x; u)\nu(dtdxdu) \mid \nu \in \tilde{\mathscr{A}}, \nu_{1,t} = P_t, 0 \le t \le 1\}$

Chapter 1
Introduction

Abstract Starting from Monge's problem in 1781, the theory of optimal mass transportation (OT for short) has been studied by many authors in many fields of research. Partly as a stochastic analog of the OT, we have been studying the so-called stochastic optimal transportation problem (SOT for short). It is a stochastic optimal control problem with fixed marginal distributions. One of the important purposes is to study the OT in the framework of the SOT. It is also a generalization of Schrödinger's problem and is related to Nelson's stochastic mechanics. We briefly describe the OT and Schrödinger's problems in such a way that one can compare the similarities between them.

1.1 Background

We are interested in the mathematical analysis of phenomena. First, we would like to find a variational problem in which a minimizer describes a phenomenon and then study the phenomenon itself via the variational problem from a mathematical point of view.

To consider a variational problem, we first construct a function S and a set A over which we minimize S. Then we find a condition under which A is not empty. Indeed, the sets under consideration in this monograph can be empty. Our problem can be described as in the following manner:

$$\inf\{S(x)|x \in A\}. \tag{1.1}$$

As a typical example, we consider the distance between two points $x_0, x_1 \in \mathbb{R}^d$ via a variational problem. Let $C([0, 1]; \mathbb{R}^d)$ and $AC([0, 1]; \mathbb{R}^d)$ denote the set of all continuous and absolutely continuous functions from $[0, 1]$ to $\mathbb{R}^d$, respectively.

$$S_{0,1}(x) := \begin{cases} \int_0^1 |\dot{x}(t)|^2 dt, & x \in AC([0, 1]; \mathbb{R}^d), \\ \infty, & x \in C([0, 1]; \mathbb{R}^d) \cap AC([0, 1]; \mathbb{R}^d)^c, \end{cases} \tag{1.2}$$

T. Mikami, *Stochastic Optimal Transportation*, SpringerBriefs in Mathematics,
https://doi.org/10.1007/978-981-16-1754-6_1

where $\dot{x}(t) := dx(t)/dt$. $S_{0,1}(x)$ can be considered a cost to move from $x(0)$ to $x(1)$ along $x = (x(t))_{0\le t\le 1}$. (1.2) means that we do not move along $x \notin AC([0, 1]; \mathbb{R}^d)$ which is too zigzag!

It seems that we move from one point to another along the line segment which connects the two points since we know that the following holds:

$$|x_1 - x_0|^2 = \inf\left\{S_{0,1}(x)|x(0) = x_0, x(1) = x_1, x \in C([0, 1]; \mathbb{R}^d)\right\}, \tag{1.3}$$

where the minimizer is

$$x_0 + t(x_1 - x_0), \quad 0 \le t \le 1. \tag{1.4}$$

$S_{0,1}(x)$ and $\{x \in C([0, 1]; \mathbb{R}^d)|x(0) = x_0, x(1) = x_1\}$ play roles of $S(x)$ and $A(\neq \emptyset)$ in (1.1), respectively.

We show that (1.3)–(1.4) hold. By Schwartz's inequality,

$$|x(1) - x(0)|^2 = \left|\int_0^1 \dot{x}(t)dt\right|^2 \le \int_0^1 |\dot{x}(t)|^2 dt, \quad x \in AC([0, 1]; \mathbb{R}^d),$$

where the equality holds if and only if $\dot{x}(t)$ is a constant, which implies (1.3)–(1.4).

We consider Euler's equation for the right-hand side of (1.3), that is, the following:

$$\text{the first variation of the right-hand side of (1.3)} = 0.$$

Let $x \in AC([0, 1]; \mathbb{R}^d)$ be a minimizer of (1.3). Then for any $y \in AC([0, 1]; \mathbb{R}^d)$ for which $y(0) = y(1) = 0$ and $\delta \in \mathbb{R}$,

$$x(0) + \delta y(0) = x_0, \quad x(1) + \delta y(1) = x_1, \quad S_{0,1}(x + \delta y) \ge S_{0,1}(x),$$

which implies that the Gâteaux derivative vanishes:

$$\lim_{\delta\to 0} \frac{S_{0,1}(x + \delta y) - S_{0,1}(x)}{\delta} = 0, \tag{1.5}$$

provided it exists. That is,

$$\frac{d}{d\delta}\int_0^1 |\dot{x}(t) + \delta\dot{y}(t)|^2 dt\bigg|_{\delta=0} = 2\int_0^1 \langle\ddot{x}(t), y(t)\rangle dt = 0, \tag{1.6}$$

since $y(0) = y(1) = 0$. Equation (1.6) implies that the following holds:

$$\ddot{x}(t) = 0, \tag{1.7}$$

which implies (1.3)–(1.4).

The above implies that there are at least two approaches to find and characterize a minimizer of (1.3). One is by inequalities on integrals. Another is by solving Euler's equation to find a critical point of (1.3).

Let $f \in C_b(\mathbb{R}^d)$, where $C_b(\mathbb{R}^d) = C_b(\mathbb{R}^d; \mathbb{R})$ denotes the set of all bounded continuous functions from $\mathbb{R}^d$ to $\mathbb{R}$. The following is an example of the so-called fixed finite-time horizon optimal control problem (see, e.g. [60, 61]): for $x_0 \in \mathbb{R}^d$,

$$\begin{aligned} &\inf\left\{S_{0,1}(x) + f(x(1))|x(0) = x_0, x \in C([0,1]; \mathbb{R}^d)\right\} \qquad (1.8)\\ &= \inf\left\{\inf\left\{S_{0,1}(x)|x(0) = x_0, x(1) = x_1, x \in C([0,1]; \mathbb{R}^d)\right\} + f(x_1)|x_1 \in \mathbb{R}^d\right\}\\ &= \inf\left\{|x_1 - x_0|^2 + f(x_1)|x_1 \in \mathbb{R}^d\right\}. \end{aligned}$$

The theory of the Hamilton–Jacobi equation is useful to study (1.8). Indeed, the last infimum in (1.8) is called the Hopf–Lax formula for the Hamilton–Jacobi equation (see Remark 2.2 in Sect. 2.1).

Equation (1.3) can be rewritten as follows:

$$|x_1 - x_0|^2 = \inf\{S_{0,1}(x) + \chi_{\{x_1\}}(x(1))|x(0) = x_0, x \in C([0,1]; \mathbb{R}^d)\}, \qquad (1.9)$$

where $\chi_A(x) := 0, x \in A, := \infty, x \notin A$. In particular, (1.3) can be considered an example of fixed finite-time horizon optimal control problems though $\chi_{\{x_1\}} \notin C_b(\mathbb{R}^d)$.

$S_{0,1}(x)$ is an action integral and (1.7) is the Newtonian Equation of motion with no external force in classical mechanics. $S_{0,1}(x)$ also plays a crucial role in R. P. Feynman's path integral formulation of quantum mechanics (see, e.g. [136]).

An action integral also plays a crucial role in the theory of small random perturbations of dynamical systems (see [64] and also [61] for a variational approach). Let $\{B(t) = B(t,\omega)\}_{t\ge 0}$ denote a d-dimensional Brownian motion on a probability space $(\Omega, \mathscr{F}, P)$. Roughly speaking, the following is known: for $x_0 \in \mathbb{R}^d$ and a Borel set $A \subset C([0,1]; \mathbb{R}^d)$,

$$P(x_0 + \varepsilon B(\cdot) \in A) \sim \exp\left(-\frac{1}{2\varepsilon^2}\inf\{S_{0,1}(x)|x \in A, x(0) = x_0\}\right), \quad \varepsilon \to 0. \qquad (1.10)$$

One intuitive interpretation of (1.10) is the following: as $\varepsilon \to 0$,

$$P(x_0 + \varepsilon B(\cdot) \in dx) \sim \exp\left(-\frac{1}{2\varepsilon^2}S_{0,1}(x)\right)dx \quad \text{on } C([0,1]; \mathbb{R}^d). \qquad (1.11)$$

The infimum in (1.10) can be rewritten as follows:

$$\inf\left\{S_{0,1}(x)+\chi_A(x)|x(0)=x_0, x\in C([0,1];\mathbb{R}^d)\right\}. \tag{1.12}$$

If $A=\{x\in C([0,1];\mathbb{R}^d)|x(1)=x_1\}$, then (1.12) is equal to (1.9).

Suppose that the set A_o of minimizers in the infimum in (1.10) exists and that the following holds: for $\delta>0$

$$\begin{aligned}&\inf\left\{S_{0,1}(x)|x\in A, x(0)=x_0\right\} \\ &<\inf\left\{S_{0,1}(x)|x\in A\cap U_\delta(A_o)^c, x(0)=x_0\right\}.\end{aligned} \tag{1.13}$$

Here

$$\begin{aligned}U_\delta(A_o)&:=\{x\in C([0,1];\mathbb{R}^d)|d(x,A_o)<\delta\},\\ d(x,A_o)&:=\inf\{\sup\{|x(t)-y(t)|:0\le t\le 1\}|y\in A_o\}.\end{aligned}$$

Then by the Laplace method,

$$P(x_0+\varepsilon B(\cdot)\in U_\delta(A_o)|x_0+\varepsilon B(\cdot)\in A)\to 1, \quad \varepsilon\to 0. \tag{1.14}$$

In particular, the analysis of a subset A_o of minimizers of $S_{0,1}$ on a set A is also crucial in the case where we only consider a rare event A. In this sense, an action integral even controls a rare event. This kind of problem also appears in the study of phase transition (see, e.g. [77]).

As a generalization of (1.3), instead of two points, we consider two sets $D_0:=\{x_{0,1},x_{0,2}\}$ and $D_1:=\{x_{1,1},x_{1,2}\}$ of distinct two points in $\mathbb{R}^d$ and study the following:

$$\begin{aligned}&\min(|x_{1,1}-x_{0,1}|+|x_{1,2}-x_{0,2}|,|x_{1,2}-x_{0,1}|+|x_{1,1}-x_{0,2}|) \\ &=\min\left\{\sum_{x(0)\in D_0}S_{0,1}(x)^{\frac{1}{2}}\middle|\{x(1)|x(0)\in D_0\}=D_1, x\in C([0,1];\mathbb{R}^d)\right\}.\end{aligned} \tag{1.15}$$

If $|x_{1,i_1}-x_{0,1}|+|x_{1,i_2}-x_{0,2}|, i_1\neq i_2$ is a minimizer, then the line segments

$$\{x_{0,1}+t(x_{1,i_1}-x_{0,1})|0\le t\le 1\} \text{ and } \{x_{0,2}+t(x_{1,i_2}-x_{0,2})|0\le t\le 1\}$$

do not intersect since the length of one side of a triangle is less than the sum of those of the other sides.

More generally, if the sets D_0 and D_1 contain more than two points and if the cardinality of D_0 and D_1 are the same, then a minimizer of the right-hand side of (1.15) gives non-intersecting line segments that connect all points of the sets D_0 and D_1. It also shows the most convenient way to move a pile of pebbles from one place to another, which gives an answer to the so-called **Monge's problem** (see [121]).

As a generalization of such problems, instead of considering general sets D_0 and D_1, assuming that $x(0)$ and $x(1)$ are random but obey given probability distributions P_0 and P_1, respectively, the **optimal transportation problem** is known. It is also called the **Monge–Kantorovich problem**, the optimal mass transportation problem, the optimal transport, the optimal transfer problem, etc. (see Sects. 1.3 and 2.1).

We consider a randomized version of (1.15). We first describe Monge's problem where $x(1)$ is assumed to be a function of $x(0)$ and the function is nonrandom. Let $\mathscr{P}(S)$ denote the set of all Borel probability measures on a topological space S. For $P_0 \in \mathscr{P}(\mathbb{R}^d)$, take an $\mathbb{R}^d$-valued random variable $X_0(\omega)$ on a probability space $(\Omega, \mathscr{F}, P)$ such that $P^{X_0} = P_0$, where P^{X_0} denotes the probability distribution of X_0. The following is a typical example of Monge's problem: for $P_1 \in \mathscr{P}(\mathbb{R}^d)$,

$$\begin{aligned} &\inf\{E[|\psi(X_0) - X_0|] \,|\, P^{\psi(X_0)} = P_1\} \qquad (1.16)\\ &= \inf\left\{\int_{\mathbb{R}^d} |\psi(x) - x| P_0(dx) \middle| P_0\psi^{-1} = P_1\right\}. \end{aligned}$$

On the left-hand side of (1.16), the roles of $S(x)$ and A in (1.1) are, respectively, played by $E[|\psi(X_0) - X_0|]$ and the set $\{\psi : \mathbb{R}^d \to \mathbb{R}^d | P^{\psi(X_0)} = P_1\}$. It is not trivial if this set is not empty.

L. V. Kantorovich considered a relaxed problem (see [83, 84]). For $P_0, P_1 \in \mathscr{P}(\mathbb{R}^d)$,

$$\begin{aligned} &\inf\{E[|X(1) - X(0)|] \,|\, P^{X(0)} = P_0, P^{X(1)} = P_1\} \qquad (1.17)\\ &= \inf\left\{E\left[\int_0^1 |\dot{X}(t)|dt\right] \middle| P^{X(0)} = P_0, P^{X(1)} = P_1, X \in AC([0,1]; \mathbb{R}^d), \text{a.s.}\right\}\\ &= \inf\{E[S_{0,1}(X)^{\frac{1}{2}}] | P^{X(0)} = P_0, P^{X(1)} = P_1, X \in AC([0,1]; \mathbb{R}^d), \text{a.s.}\}\\ &= \inf\{E[S_{0,1}(X)^{\frac{1}{2}}] + \chi_{\{P_1\}}(P^{X(1)}) | P^{X(0)} = P_0, X \in AC([0,1]; \mathbb{R}^d), \text{a.s.}\}. \end{aligned}$$

Here $X = \{X(t)\}_{0 \le t \le 1}$ denotes a stochastic process defined on a possibly different probability space. When it is not confusing, we use the same notation P for different probability measures. In a sense, (1.17) can be considered a fixed finite-time horizon **stochastic** optimal control problem (see, e.g. [60, 61, 92]). Indeed,

$$\chi_{\{P_1\}}(P^{X(1)}) = \sup\left\{E[f(X(1))] - \int_{\mathbb{R}^d} f(x)P_1(dx) \middle| f \in C_b(\mathbb{R}^d)\right\}.$$

If $P^{X(t)} = \delta_{\{x_t\}}, t = 0, 1$, then (1.17) implies (1.3), where $\delta_{\{x\}}$ denotes the delta measure on $\{x\}$.

On the left-hand side of (1.17), the roles of $S(x)$ and A in (1.1) are, respectively, played by $E[|X(1) - X(0)|]$ and the set of all $\mathbb{R}^d \times \mathbb{R}^d$-valued random variables $(X(0), X(1))$ defined on a probability space with $P^{X(i)} = P_i, i = 0, 1$.

In the third infimum of (1.17), the roles of $S(x)$ and A in (1.1) are, respectively, played by $E[S_{0,1}(X)^{\frac{1}{2}}]$ and the set of all absolutely continuous stochastic processes defined on a possibly different probability space with probability distributions at time t are $P_t, t = 0, 1$. This set is not empty since for random variables $X(0), X(1)$, the following function is absolutely continuous, a.s.:

$$[0, 1] \ni t \mapsto X(0) + t(X(1) - X(0)) \in \mathbb{R}^d.$$

The existence of an absolutely continuous stochastic process with given probability distributions at all times $t \in [0, 1]$ is not trivial (see [3, 107] and also Sect. 3.2.2). The existence of a continuous stochastic process with given probability distributions at all times $t \in [0, 1]$ is also given in [15].

On the right-hand side of (1.17), (1.7) holds for a minimizer $\{X(t, \omega)\}_{0 \le t \le 1}$ defined on a probability space $(\Omega, \mathscr{F}, P)$. It is also known that under an appropriate condition, there exists a nonrandom function ψ such that

$$X(t) = X(0) + t(\psi(X(0)) - X(0)), \quad 0 \le t \le 1, \text{a.s.}.$$

That is, (1.17) coincides with (1.16) and ψ connects almost all points of $\{X(0, \omega)|\omega \in \Omega\}$ and $\{X(1, \omega)|\omega \in \Omega\}$ (see Sect. 1.3 for more discussion).

It is also known that there exist different kinds of useful functions which connect almost all points of $\{X(0, \omega)|\omega \in \Omega\}$ and $\{X(1, \omega)|\omega \in \Omega\}$ (see, e.g. [46, 87, 130, 142, 152]).

An absolutely continuous stochastic process $\{X_0 + t(\psi(X_0) - X_0)\}_{0 \le t \le 1}$ is deterministic, in that

$$\sigma[X_0 + s(\psi(X_0) - X_0); 0 \le s \le t] = \sigma[X_0], \quad 0 \le t \le 1,$$

where $\sigma[X_0]$ denotes the σ-field generated by X_0. In (1.17), it is not necessarily true for all X, though we expect it is for an optimal X in the case where (1.17) coincides with (1.16).

In (1.17), we did not consider a random external force. Our interest is to consider its generalizations for semimartingales.

We describe an example. Let $\{U(t) = U(t, \omega)\}_{t \ge 0}$ and $X_0 = X_0(\omega)$ be an $\mathbb{R}^d$-valued absolutely continuous stochastic process such that $U(0) = 0$ and random variable on $(\Omega, \mathscr{F}, P)$, respectively. For $\varepsilon \ge 0$,

$$X^{\varepsilon,U}(t) = X^{\varepsilon,U}(t, \omega) := X_0(\omega) + U(t, \omega) + \varepsilon B(t, \omega), \quad 0 \le t \le 1. \tag{1.18}$$

Here $B(t)$ plays a role of random external force which is uncontrollable and we would like to find a nice $U(t)$. We also cheat the measurability of $U(t, \omega)$ to a filtration. Notice that $X^{0,U}$ is absolutely continuous, a.s.

Let $f \in C_b(\mathbb{R}^d)$. The following is a typical example of the so-called finite-time horizon stochastic optimal control problem (see, e.g. [60, 61, 92]):

$$\inf\left\{E\left[S_{0,1}(U) + f(X^{\varepsilon,U}(1))\right] \middle| U \in AC([0,1];\mathbb{R}^d), \text{a.s.}\right\} \tag{1.19}$$
$$= \inf\left\{\inf\left\{E\left[S_{0,1}(U)\right] \middle| P^{X^{\varepsilon,U}(1)} = P, U \in AC([0,1];\mathbb{R}^d), \text{a.s.}\right\}\right.$$
$$\left. + \int_{\mathbb{R}^d} f(x)P(dx) \middle| P \in \mathcal{P}(\mathbb{R}^d)\right\}.$$

We recall that $P^{X^{\varepsilon,U}(0)} = P^{X_0}$ is fixed. The theory of the Hamilton–Jacobi–Bellman equation is useful to study (1.19).

As an analogy of (1.17), we consider the following: for $P_1 \in \mathcal{P}(\mathbb{R}^d)$,

$$\inf\left\{E\left[S_{0,1}(U)\right] \middle| P^{X^{\varepsilon,U}(1)} = P_1, U \in AC([0,1];\mathbb{R}^d), \text{a.s.}\right\} \tag{1.20}$$
$$= \inf\left\{E\left[S_{0,1}(U)\right] + \chi_{\{P_1\}}\left(P^{X^{\varepsilon,U}(1)}\right) \middle| U \in AC([0,1];\mathbb{R}^d), \text{a.s.}\right\}.$$

$E\left[S_{0,1}(U)\right]$ and $\{U|P^{X^{\varepsilon,U}(1)} = P_1\}$ play roles of $S(x)$ and A in (1.1), respectively. We expect that by the theory of semimartingales, letting $\varepsilon \to 0$, the study of (1.20) with $S_{0,1}(U)$ replaced by $\int_0^1 |\dot{U}(t)|dt$ gives more information on (1.17).

More generally, we consider the following: for $A \subset \mathcal{P}(C([0,1];\mathbb{R}^d))$,

$$\inf\left\{E\left[S_{0,1}(U)\right] \middle| P^{X^{\varepsilon,U}} \in A, U \in AC([0,1];\mathbb{R}^d), \text{a.s.}\right\} \tag{1.21}$$
$$= \inf\left\{E\left[S_{0,1}(U)\right] + \chi_A(P^{X^{\varepsilon,U}}) \middle| U \in AC([0,1];\mathbb{R}^d), \text{a.s.}\right\}.$$

It is not trivial if the set $\{U|P^{X^{\varepsilon,U}} \in A\}$ is not empty and if $E\left[S_{0,1}(U)\right]$ is finite at least in one point of the set $\{U|P^{X^{\varepsilon,U}} \in A\}$ (see Sects. 2.2.3 and 3.1.2 for more discussion).

Equation (1.21) is equal to (1.20) if

$$A = \{P \in \mathcal{P}(C([0,1];\mathbb{R}^d))|P\pi(1)^{-1} = P_1\}, \tag{1.22}$$

where

$$\pi(t)(\omega) := \omega(t), \quad 0 \le t \le 1, \omega \in C([0,1];\mathbb{R}^d).$$

An optimal control problem for stochastic processes is called a stochastic optimal control problem. By the **stochastic optimal transportation problem**, we denote an optimal transportation problem for stochastic processes. The problem discussed above is a typical example. In particular, (1.20) is an example of the so-called **Schrödinger's problem**. Euler's equation for Schrödinger's problem is the so-called **Schrödinger's functional equation** (see [140, 141], also [96] for a nice survey, and Definition 1.4 and Proposition 3.6 for more discussion). The continuum limit of Schrödinger's problem is the so-called **Nelson's problem** that is also a class of the SOTs, where in (1.21), for $\{P_t\}_{0\le t\le 1} \subset \mathscr{P}(\mathbb{R}^d)$,

$$A = \{P \in \mathscr{P}(C([0,1];\mathbb{R}^d)) | P\pi(t)^{-1} = P_t, 0 \le t \le 1\}. \tag{1.23}$$

If the set A in (1.23) is not empty, then we say that the **superposition principle** for $\{P_t\}_{0\le t\le 1}$ holds (see [125, 126] and also Sects. 1.4, 3.1, and 3.2 for more discussion).

Remark 1.1 The terminologies in the SOT are not necessarily unique since it is a rapidly developing research field. The stochastic optimal transportation problem is also called the semimartingale optimal transportation problem. Since the SOT is different from the so-called martingale optimal transportation problem, we prefer to use the terminology "stochastic optimal transportation problem" to avoid confusion. Schrödinger's problem is also called the Schrödinger bridge problem these days. Schrödinger's functional equation is also called the Schrödinger system. Since the Schrödinger system also means a system of Schrödinger's PDEs, we prefer to use the terminology "Schrödinger's functional equation" to avoid confusion.

1.2 Motivation

Suppose that we investigate the statistical and dynamical properties of a random particle on the time interval [0, 1] from the probability distributions of the particle at times $t = 0$ and $t = 1$.

One question is if the random particle is described by a stochastic process. If it is, then the candidate can be a Bernstein process (see Definition 1.5 in Sect. 1.4 and also [12, 79]). We would also like to find a random differential equation that is satisfied by the stochastic process (see [39, 90, 91, 149] and the references therein). Here we point out that the differential equation does not necessarily determine the statistical property of the solution (see, e.g. [137]).

The following question is if the stochastic process is also a Markov process that satisfies a stochastic differential equation (SDE for short). Here we point out that a Markov process is a Bernstein process, but not vice versa (see, e.g. [79]).

One possible approach to these questions is a stochastic optimal control for stochastic processes (see, e.g. [60, 61, 92]). Indeed, if we know an action integral for each sample path and if we can find a random optimal path by the principle of least action applied to the mean value of an action integral, then the optimal path might explain some aspects of the random particle. Since the probability distributions at times $t = 0$ and $t = 1$ are fixed, the problem is not a standard stochastic optimal control. This kind of mathematical problem is SOT. Schrödinger's problem is a typical example (see Sect. 1.4). If the particle does not move randomly, then the SOT can be considered the OT (see Sect. 1.3).

Our interest is the construction of the Markov process which describes some aspect of the random phenomenon, from given partial information, including marginal distributions. The SOT is one approach to this and is a class of marginal problems, in that it constructs a semimartingale as a minimizer of a variational problem with given marginal distributions.

The Brunn–Minkowski inequality plays a crucial role in geometric analysis and implies an isoperimetric inequality (see [87] and also [30, 139]). It can be proved by the so-called Knothe–Rosenblatt rearrangement (see [87]) which was also used to study the transportation cost inequality (see [146]) and to study the log-Sobolev inequality (see [16, 17]).

The Knothe–Rosenblatt rearrangement can be defined inductively by the minimizer of the one-dimensional OT with a strictly convex cost for conditional distributions (see [30, 87] and also [114] for the SOT analog). The minimizer of the one-dimensional OT with a strictly convex cost can be also found in [46, 99, 134] (see also [124, 130, 142, 152, 153]).

In information theory, an inequality on the Renyi entropy power implies the entropy power inequality. It is interesting to see that the inequality also implies the Brunn–Minkowski inequality (see, e.g. [41, section 17.8]). We also refer readers to [37, 44, 129] on Sinkhorn's algorithm as a related topic of Schrödinger's problem (see also [127]).

It seems that the OT, the SOT, geometric inequalities, functional inequalities, and information theory are strongly related.

We would also like to refer to the Euclidean quantum mechanics as a possible application of the SOT (see [2, 38, 43, 157, 158] and the references therein).

We would like to develop the OT in the framework of the SOT and find applications of the SOT in many fields in the future.

1.3 Optimal Transportation Problem

In this section, we describe the optimal mass transportation problem from the viewpoint of stochastic mechanics.

In 1781, G. Monge proposed the following problem at the Paris Royal Academy of Sciences (see [121]):

What is the most convenient way to move a pile of pebbles from one place to another?

In the twentieth century, this problem was generalized by L. V. Kantorovich (see [83, 84]) so that one can consider it in a mathematically easier framework and is called the optimal mass transportation problem or the Monge–Kantorovich problem nowadays (see, e.g. [54, 130, 152, 153]). Many kinds of studies on the OT have been done rapidly, e.g. the applications of the OT to partial differential equations, limit theorems and the log-Sobolev inequality for probability measures, the geometry of the infinite dimensional space, economics and image processing and the OT over the Riemannian manifold and Wiener space (see, e.g. [25, 47, 50, 58, 81, 98, 148] and the references therein).

We explain the mathematical idea of the OT as the problem of stochastic mechanics determined by the principle of least action. Take two sets S_0 and S_1 which consist of finite distinct n_0 and n_1 points in $\mathbb{R}^d$, respectively:

$$S_0 := \{x_1, \cdots, x_{n_0}\}, \quad S_1 := \{y_1, \cdots, y_{n_1}\} \subset \mathbb{R}^d.$$

Let $N_0(i)$ and $N_1(j)$ denote the number of pebbles at $x_i \in S_0$ and the number of pebbles one would like to move from S_0 to $y_j \in S_1$, respectively. Then

$$N := \sum_{i=1}^{n_0} N_0(i) = \sum_{j=1}^{n_1} N_1(j). \tag{1.24}$$

Here we suppose that $N_0(i), N_1(j) \geq 1$. We also suppose that we move pebbles from the same place in S_0 to the same place in S_1. Then $n_0 \geq n_1$ and for any $j \in \{1, \cdots, n_1\}$, there exists $k(j) \geq 1$ and $i_{j,1} < \cdots < i_{j,k(j)}$ such that

$$N_1(j) = \sum_{k=1}^{k(j)} N_0(i_{j,k}). \tag{1.25}$$

Suppose that the cost to move a pebble from x to y is proportional to $|y - x|$. Let $\psi : S_0 \longrightarrow S_1$ be a function which explains how to move N pebbles from S_0 to S_1. Then the total cost to move pebbles from S_0 to S_1 by ψ is proportional to

$$\sum_{i=1}^{n_0} |\psi(x_i) - x_i| N_0(i).$$

To minimize the cost, we consider the following problem under (1.24):

$$\inf\left\{\sum_{i=1}^{n_0} |\psi(x_i) - x_i| N_0(i) \,\middle|\, \sum_{i: x_i \in \psi^{-1}(y_j)} N_0(i) = N_1(j),\ j \in \{1, \cdots, n_1\}\right\} \tag{1.26}$$

$$= N \inf\left\{\int_{\mathbb{R}^d} |\psi(x) - x| \left(\sum_{i=1}^{n_0} \frac{N_0(i)}{N} \delta_{x_i}(dx)\right) \,\middle|\, \left(\sum_{i=1}^{n_0} \frac{N_0(i)}{N} \delta_{x_i}(dx)\right) \psi^{-1} = \sum_{j=1}^{n_1} \frac{N_1(j)}{N} \delta_{y_j}(dy)\right\}.$$

Here δ_{x_i} and $P\psi^{-1}$ denote a delta measure on $\{x_i\}$ and an image measure or a push forward of a probability measure P by ψ, respectively.

From the example above, we obtain the following.

Definition 1.1 (Monge's Problem) Let $P_0, P_1 \in \mathscr{P}(\mathbb{R}^d)$ and $c : \mathbb{R}^d \times \mathbb{R}^d \longrightarrow \mathbb{R}$ be Borel measurable. Find a sufficient condition under which there exists a unique minimizer ψ of the following:

$$T_M(P_0, P_1) := \inf\left\{ \int_{\mathbb{R}^d} c(x, \psi(x)) P_0(dx) \middle| P_0\psi^{-1} = P_1 \right\}. \tag{1.27}$$

In Monge's problem, the integral in the infimum is not linear in ψ. We describe Kantorovich's approach to simplify the problem.

$$\begin{aligned} T_M(P_0, P_1) &\geq T(P_0, P_1) \qquad (1.28)\\ &:= \inf\left\{ \int_{\mathbb{R}^d \times \mathbb{R}^d} c(x, y)\mu(dxdy) \middle| \mu_1 = P_0, \mu_2 = P_1 \right\}, \end{aligned}$$

where for $\mu \in \mathscr{P}(\mathbb{R}^d \times \mathbb{R}^d)$, $\mu_1(dx) := \mu(dx \times \mathbb{R}^d)$, $\mu_2(dy) := \mu(\mathbb{R}^d \times dy)$.

This leads us to Kantorovich's idea.

Definition 1.2 (Optimal Transportation Problem/Monge–Kantorovich Problem) Let $P_0, P_1 \in \mathscr{P}(\mathbb{R}^d)$ and $c : \mathbb{R}^d \times \mathbb{R}^d \longrightarrow \mathbb{R}$ be Borel measurable. Find a sufficient condition under which there exists a unique minimizer μ of $T(P_0, P_1)$ and a Borel measurable function $\psi : \mathbb{R}^d \longrightarrow \mathbb{R}^d$ such that the following holds:

$$\mu(dxdy) = P_0(Id, \psi)^{-1}(dxdy)(= P_0(dx)\delta_{\psi(x)}(dy)), \tag{1.29}$$

where Id denotes an identity mapping, i.e. $Id(x) = x$.

If (1.29) is true, then $T_M(P_0, P_1) = T(P_0, P_1)$ and ψ in (1.29) is the unique minimizer of $T_M(P_0, P_1)$.

Remark 1.2 If P_0 and $P_1 \in \mathscr{P}(\mathbb{R}^d)$ are compactly supported, and if P_0 does not have a point mass, then $T_M(P_0, P_1) = T(P_0, P_1)$ (see [139, section 1.5, Theorem 1.33]).

In classical mechanics, dynamics are characterized by a given Lagrangian function

$$L(\cdot, \cdot; \cdot) : [0, 1] \times \mathbb{R}^d \times \mathbb{R}^d \longrightarrow \mathbb{R}.$$

The motion of the system is determined by the principle of least action. More precisely, consider the following minimization problem: for any points $x_0, x_1 \in \mathbb{R}^d$,

$$I(x_0, x_1) := \inf\left\{ \int_0^1 L(t, x(t); \dot{x}(t))dt \,\middle|\, x(\cdot) \in AC([0,1]; \mathbb{R}^d), \right. \tag{1.30}$$
$$\left. x(t) = x_t, t = 0, 1 \right\},$$

A graph of a function $x(t)$ which minimizes $I(x_0, x_1)$ determines the path of a classical mechanical motion which connects x_0 and x_1. Here a path of $x(t)$ is not necessarily unique.

In the case where $c = I$,

$$I(x_0, x_1) = T(\delta_{x_0}, \delta_{x_1})$$

(see (1.28)). Besides, in the case where $L = L(u)$ and L is convex, the following holds:

$$T(P_0, P_1) = \inf\{E[L(X(1) - X(0))] | P^{X(t)} = P_t, t = 0, 1\} \tag{1.31}$$
$$= \inf\left\{ E\left[\int_0^1 L(\dot{X}(t))dt\right] \middle| X \in AC([0,1]; \mathbb{R}^d), \text{ a.s., } P^{X(t)} = P_t, t = 0, 1 \right\}.$$

Equation (1.31) implies that the OT is a generalization of (1.30). Equation (1.31) is true since, by Jensen's inequality,

$$I(x_0, x_1) = L(x_1 - x_0), \tag{1.32}$$

and $x_0 + t(x_1 - x_0)$ is a (unique if L is strictly convex) minimizer of (1.30). If $L(u) = |u|^2$, P_0 has a density function $p_0(x) := P_0(dx)/dx$ and P_0, P_1 have finite second moments, then there exists a convex function φ such that the minimizer of $T(P_0, P_1)$ is $P_0(dx)\delta_{D\varphi(x)}(dy)$ (see [21, 22]). In addition, if $p_1(x) := P_1(dx)/dx$ exists, then we have the following Monge–Ampère equation:

$$p_0(x) = p_1(D\varphi(x))det(D^2\varphi(x)), \quad x \in \mathbb{R}^d. \tag{1.33}$$

More precisely, for any $f \in C^2(\mathbb{R}^d)$,

$$\int_{\mathbb{R}^d} f(D\varphi(x))p_0(x)dx = \int_{\mathbb{R}^d} f(x)p_1(x)dx. \tag{1.34}$$

In application, we also have to consider the case where $u \mapsto L(u)$ is concave. Even in this case, $T(P_0, P_1)$ can be finite if we suppose an appropriate integrability condition on P_0 and P_1 (see, e.g. [124, 130, 142]).

1.4 Schrödinger's Problem

In this section, we describe Schrödinger's problem from the viewpoint of stochastic optimal control theory (see [140, 141]). As a nice survey on Schrödinger's problem and the OT, we refer readers to [96]. We also refer them to [155] on Schrödinger's problem.

E. Schrödinger obtained the so-called Schrödinger's Functional Equation (Func. Eqn. for short) or the Schrödinger system as Euler's equation for his variational problem (see [13, 36, 63, 119, 135] and the references therein). Schrödinger's problem was mathematically formulated by S. Bernstein who defined a Bernstein process which is also called a reciprocal process nowadays (see [12, 79] and the references therein). The theory of the SDE for reciprocal processes was given by B. Jamison (see [80]). Schrödinger's idea also brought about J. C. Zambrini's stochastic control approach to the SOT and E. Nelson's stochastic mechanics (see [45, 62, 125, 126, 156] and the references therein). Nelson's stochastic mechanics proposed the mathematical problem of the construction of a Markov diffusion process from the Fokker–Planck equation (see [19, 28, 29, 31–35, 59, 103, 111, 112, 123, 150, 159]). In this sense both the OT and Schrödinger's problem can be considered important classes of the so-called marginal problems. Schrödinger's problem is also related to the theory of large deviations, entropic estimates, and functional inequalities (see, e.g. [1, 7, 39, 40, 51, 53, 57, 69, 101, 128, 131, 154] and the references therein).

We describe Schrödinger's problem, which is similar to Monge's. Suppose that there are N particles, in S_0, which move independently (in the probability sense), with a given transition probability, to S_1. Here we use the same notation as in (1.24) for the numbers of particles in S_0 and S_1 and assume that (1.24) holds, but not (1.25). We would like to know the maximal probability of such events. Let $g_{ij} > 0$ and N_{ij} denote the given transition probability that a particle at $x_i \in S_0$ moves to $y_j \in S_1$ and the number of particles which move from $x_i \in S_0$ to $y_j \in S_1$, respectively. Since each particle is independent, the probability of such an event is given by the following:

$$\prod_{i=1}^{n_0} \frac{N_0(i)!}{N_{i1}! \times \cdots \times N_{in_1}!} g_{i1}^{N_{i1}} \times \cdots \times g_{in_1}^{N_{in_1}}, \tag{1.35}$$

under the following constraints: for $i \in \{1, \cdots, n_0\}$, $j \in \{1, \cdots, n_1\}$,

$$\sum_{\ell=1}^{n_1} N_{i\ell} = N_0(i), \quad \sum_{k=1}^{n_0} N_{kj} = N_1(j), \quad \sum_{k=1}^{n_0} \sum_{\ell=1}^{n_1} N_{k\ell} = N. \tag{1.36}$$

Recall that $N_0(i)$ and $N_1(j)$ are fixed numbers. Suppose that each N_{ij} is sufficiently large so that by Stirling's formula,

$$\log(N_{ij}!) \sim N_{ij}(\log N_{ij} - 1) + O(\log N_{ij}) \sim N_{ij}(\log N_{ij} - 1 + o(1)).$$

Then, to maximize the probability, considering the logarithm of the probability in (1.35), we consider the following problem:

$$\max\left\{\sum_{i=1}^{n_0}\sum_{j=1}^{n_1}(N_{ij}\log g_{ij} - N_{ij}\log N_{ij}) : \right. \tag{1.37}$$

$$\left.\sum_{\ell=1}^{n_1} N_{i\ell} = N_0(i), \sum_{k=1}^{n_0} N_{kj} = N_1(j), 1 \le i \le n_0, 1 \le j \le n_1\right\}.$$

We define discrete probabilities $\mu := (\mu_i)_{i=1}^{n_0}$, $\nu := (\nu_j)_{j=1}^{n_1}$ and $(p_{ij})_{1\le i\le n_0, 1\le j\le n_1}$ by the following:

$$\mu_i := \frac{N_0(i)}{N}, \quad \nu_j := \frac{N_1(j)}{N}, \quad p_{ij} := \frac{N_{ij}}{N}.$$

Then the maximizer of (1.37) is the minimizer of the following:

$$\mathrm{v}(\mu,\nu) := \min\left\{\sum_{i=1}^{n_0}\sum_{j=1}^{n_1} p_{ij}\log\frac{p_{ij}}{\mu_i g_{ij}} : \right. \tag{1.38}$$

$$\left.\sum_{\ell=1}^{n_1} p_{i\ell} = \mu_i, \sum_{k=1}^{n_0} p_{kj} = \nu_j, 1 \le i \le n_0, 1 \le j \le n_1\right\}.$$

For a topological space S and $P, Q \in \mathcal{P}(S)$, let $H(P|Q)$ denote the relative entropy of P with respect to Q:

$$H(P|Q) := \begin{cases} \displaystyle\int_S \log\frac{P(dx)}{Q(dx)} P(dx), & P \ll Q, \\ \infty, & \text{otherwise.} \end{cases} \tag{1.39}$$

Equation (1.38) leads us to consider Schrödinger's problem the minimization problem of a relative entropy with constraints on marginal distributions.

Definition 1.3 (Schrödinger's Problem) Let S be a topological space, $\mathcal{P}(S)$ be endowed with a weak topology and $S \ni x \mapsto q(x, dy) \in \mathcal{P}(S)$ be measurable. Study the minimizer of the following: for $P_0, P_1 \in \mathcal{P}(S)$,

$$V(P_0, P_1) := \inf\{H(\mu(dxdy)|P_0(dx)q(x,dy)) : \mu_1 = P_0, \mu_2 = P_1\} \tag{1.40}$$

(see (1.28) for notation). In this monograph, we also call the infimum itself in (1.40) Schrödinger's problem for simplicity when it is not confusing.

E. Schrödinger obtained the so-called Schrödinger's Func. Eqn. by the continuum limit of Euler's equation to (1.38). We state it in a more general setting.

Definition 1.4 (Schrödinger's Func. Eqn.) Let S be a topological space and q be a positive continuous function on $S \times S$. For $\mu_1, \mu_2 \in \mathscr{P}(S)$, find a product measure $\nu_1(dx_1)\nu_2(dx_2)$ of nonnegative σ-finite Borel measures on S for which the following holds:

$$\begin{cases} \mu_1(dx_1) = \nu_1(dx_1) \int_S q(x_1, x_2)\nu_2(dx_2), \\ \mu_2(dx_2) = \nu_2(dx_2) \int_S q(x_1, x_2)\nu_1(dx_1). \end{cases} \tag{1.41}$$

Remark 1.3 Equation (1.41) is equivalent to the following:

$$\begin{cases} \mu_1(dx_1) = \nu_1(dx_1) \int_S q(x_1, x_2) \left(\int_S q(x, x_2)\nu_1(dx) \right)^{-1} \mu_2(dx_2), \\ \nu_2(dx_2) = \left(\int_S q(x, x_2)\nu_1(dx) \right)^{-1} \mu_2(dx_2). \end{cases} \tag{1.42}$$

If one finds a solution ν_1 in the first equation of (1.42) and defines ν_2 by the second, then $\nu_1(dx_1)\nu_2(dx_2)$ satisfies (1.41). In particular, one does not have to solve a system of equations, but only one (see Definition 3.7 and Proposition 3.6 in Sect. 3.2.1). The first equation of (1.42) can be also considered the SOT version of the Monge–Ampère equation (1.33) in the OT (see Remark 2.10 in Sect. 2.3). Schrödinger's Func. Eqn. is also called the Schrödinger system, which also means a system of Schrödinger's equations in the PDE context. To avoid confusion, we use the terminology "Schrödinger's Func. Eqn." which was introduced to us by a mathematical physicist more than 30 years ago.

S. Bernstein generalized Schrödinger's idea and introduced the so-called Bernstein process which is also called a reciprocal process (see, e.g. [12, 79]).

Definition 1.5 (Bernstein Process) An $\mathbb{R}^d$-valued stochastic process $\{X(t)\}_{0 \le t \le 1}$ defined on a probability space $(\Omega, \mathscr{F}, P)$ is called a Bernstein process or a reciprocal process if the following holds: for s, t, u such that $0 \le s < t < u \le 1$,

$$P(X(t) \in dx | X(v), v \in [0, s] \cup [u, 1]) = P(X(t) \in dx | X(s), X(u)). \tag{1.43}$$

B. Jamison gave a characterization of a reciprocal process (see [79]).

Theorem 1.1 *Let $p(s, x; t, y)$ be a positive transition probability density of a Markov process, $0 \le s < t \le 1, x, y \in \mathbb{R}^d$. For any $\mu \in \mathscr{P}(\mathbb{R}^d \times \mathbb{R}^d)$, the Bernstein process $\{X(t)\}_{0 \le t \le 1}$ which satisfies the following exists and is unique: for s, t, u such that $0 \le s < t < u \le 1$,*

$$P(X(t) \in dx | X(s), X(u)) = \frac{p(s, X(s); t, x) p(t, x; u, X(u))}{p(s, X(s); u, X(u))} dx, \quad (1.44)$$
$$P^{(X(0), X(1))} = \mu.$$

In particular, $\{X(t)\}_{0 \le t \le 1}$ is Markovian, i.e., for s, t such that $0 \le s < t \le 1$,

$$P(X(t) \in dx | X(v), v \in [0, s]) = P(X(t) \in dx | X(s)) \quad (1.45)$$

if and only if there exist σ-finite measures ν_0, ν_1 such that the following holds:

$$\mu(dxdy) = \nu_0(dx) p(0, x; 1, y) \nu_1(dy). \quad (1.46)$$

R. Fortet gave the first result for Schrödinger's Func. Eqn. (see [63]). A. Beurling solved Schrödinger's Func. Eqn. on a compact topological space (see [13]). B. Jamison also solved it on a σ-compact metric space (see [79] and also [36]). The theory of the SDE for Schrödinger's Func. Eqn. was given by B. Jamison (see [80]). The solution is Doob's h-path process (see [49]) with given two endpoint marginals.

$$g(t, x) := \frac{1}{\sqrt{2\pi t}^d} \exp\left(\frac{|x|^2}{2t}\right), \quad t > 0, x \in \mathbb{R}^d. \quad (1.47)$$

We describe the construction, by Schrödinger's Func. Eqn., of an h-path process for Brownian motion on the time interval [0, 1] with given initial and terminal distributions.

Theorem 1.2 (See [80]) *Let $P_0, P_1 \in \mathscr{P}(\mathbb{R}^d)$. If $p(1, x) := P_1(dx)/dx$ exists, then there uniquely exists the h-path process $\{X_h(t)\}_{0 \le t \le 1}$ for Brownian motion on the time interval $[0, 1]$ with initial and terminal distributions P_0 and P_1, respectively. That is, the following holds weakly:*

$$dX_h(t) = b_h(t, X_h(t)) dt + dW(t), \quad 0 < t < 1, \quad (1.48)$$
$$P^{X_h(t)} = P_t, \quad t = 0, 1.$$

In particular, the following holds:

$$P^{(X_h(0), X_h(1))}(dxdy) = \nu_1(dx) g(1, x - y) \nu_2(dy). \quad (1.49)$$

Here $W(\cdot)$ denotes a d-dimensional Brownian motion (see, e.g. [75]) and

$$b_h(t,x) := D_x \log\left(\int_{\mathbb{R}^d} g(1-t, x-y)\nu_2(dy)\right), \tag{1.50}$$

where $\nu_i, i = 1, 2$ is a solution to (1.41) with $\mu_{i+1} = P_i$, $q = g(1, x-y)$ and $D_x := (\partial/\partial x_i)_{i=1}^d$.

Remark 1.4 Equation (1.49) implies that X_h is a Markovian Bernstein process with $p(s, x; t, y) = g(t-s, y-x)$. That is, Schrödinger's Func. Eqn. can be solved by constructing the h-path process with a given transition probability density of a Markov process and with given initial and terminal distributions. $P^{(X_h(0), X_h(1))}$ is absolutely continuous with respect to $P_0 \times P_1$. Indeed, from (1.41) and (1.49),

$$P^{((X_h(0), X_h(1))}(dxdy) = P_0(dx)P_1(dy)g(1, y-x)\exp(-v(1, y) - u(0, x)), \tag{1.51}$$

where

$$u(0,x) : = \log\left(\int_{\mathbb{R}^d} g(1, x-y)\nu_2(dy)\right), \tag{1.52}$$
$$v(1,y) : = \log\left(\int_{\mathbb{R}^d} g(1, x-y)\nu_1(dx)\right).$$

Let $\{\beta(t)\}_{0\le t\le 1}$ and $\{W(t)\}_{0\le t\le 1}$ denote a progressively measurable $\mathbb{R}^d$-valued stochastic process and a d-dimensional Wiener process defined on the same complete filtered probability space, respectively (see, e.g. [61]). Consider the following:

$$dX^\beta(t) = \beta(t)dt + dW(t), \quad 0 < t < 1, \tag{1.53}$$
$$P^{X^\beta(0)} = P_0.$$

If $q(x, dy) = g(1, y-x)dy$ in (1.40), then it is known that the following holds:

$$V(P_0, P_1) = \inf\{H(P^{X^\beta})|P^{X^0})|P^{X^\beta(t)} = P_t, t = 0, 1\}, \tag{1.54}$$

$$H(P^{X^\beta}|P^{X^0}) = E\left[\int_0^1 \frac{1}{2}|\beta(t)|^2 dt\right] \tag{1.55}$$

(see [135]). In particular, the minimizer of (1.54) is the h-path process in Theorem 1.2, provided it is finite (see [156]). This implies that Schrödinger's problem is a stochastic control problem with fixed initial and terminal distributions in (1.54) and that the minimizer solves Schrödinger's Func. Eqn. We consider a generalization of this problem and call it a **stochastic optimal transportation problem** (SOT for short).

Suppose that in (1.35), there exist $\varepsilon > 0$, $c(i, j) \geq 0$ such that

$$g_{ij} = g_{ij}^{\varepsilon} = \left(\sum_{\ell=1}^{n_1} \exp\left(-\frac{c(i,\ell)}{\varepsilon}\right)\right)^{-1} \exp\left(-\frac{c(i,j)}{\varepsilon}\right), \quad 1 \leq i \leq n_0, 1 \leq j \leq n_1.$$

Then $\mathrm{v}(\mu, \nu) = \mathrm{v}_{\varepsilon}(\mu, \nu)$ satisfies the following:

$$\begin{aligned} &\lim_{\varepsilon \to 0} \varepsilon \mathrm{v}_{\varepsilon}(\mu, \nu) \qquad (1.56) \\ &= \min\left\{\sum_{i=1}^{n_0}\sum_{j=1}^{n_1} p_{ij}c(i,j) \,\middle|\, \sum_{j=1}^{n_1} p_{ij} = \mu_i, \sum_{i=1}^{n_0} p_{ij} = \nu_j, 1 \leq i \leq n_0, 1 \leq j \leq n_1, \right. \\ &\qquad \left. \sum_{k=1}^{n_0}\sum_{\ell=1}^{n_1} p_{k\ell} = 1\right\} - \sum_{i=1}^{n_0} \mu_i \min\{c(i,j) | 1 \leq j \leq n_1\}. \end{aligned}$$

This can be considered an example of discrete OTs (see [95, 108, 120]).

The discussion above implies that a class of the OTs is the zero-noise limit of Schrödinger's problem and its generalizations.

In Chap. 2, first, we briefly describe the Duality Theorem for the OT. Then we introduce two classes of the SOTs, with a convex cost function, which are the stochastic control problems for semimartingales with given marginals at two endpoints and at all times, respectively. The first case where two endpoint marginals are fixed is a generalization of Schrödinger's problem. The second case where marginals at all times are fixed can be considered a continuum limit of the first one and can be considered a generalization of Nelson's problem. Indeed, in Nelson's problem, the coefficient of the first-order differential operator in the Fokker–Planck equation can be written as $D_x\psi(t, x)$ for some function $\psi(t, x)$. It turned out that it is the minimizer of a stochastic control problem in the case where marginals at all times are fixed (see [103, Proposition 3.1]). We also give the Duality Theorems for the SOTs. In the case where two endpoint marginals are fixed, we show that the minimizer of the SOT satisfies the forward–backward SDE (FBSDE for short), give a sufficient condition for the minimum in the SOT to be finite, and discuss the zero-noise limit of the SOT. In particular, we give the probabilistic proof of Monge's problem with a quadratic cost and the Duality Theorem for the OT. Our proof of the Duality Theorems for the SOTs are revised from our originally published papers thanks to D. Trevisan's recent result on the superposition principle (see [150]) and will appear in our recent paper [118].

In Chap. 3, we consider the finiteness, the semiconcavity, and the continuity of the minimum in Schrödinger's problem and the regularity of the solution of Schrödinger's Func. Eqn. We also discuss the construction of stochastic processes with given marginal distributions and an application of the OT and the superposition principle in [19] to the SOT with a nonconvex cost function in the one-dimensional case. Here [19] is a generalization of [150]. We point out that the superposition principle can be also considered a class of marginal problems.

In this monograph, we do not discuss the Knothe–Rosenblatt rearrangement and its stochastic analog (see Sect. 1.2).

Chapter 2
Stochastic Optimal Transportation Problem

Abstract We briefly describe the OT and then discuss two classes of the SOTs where two endpoint marginal distributions or marginal distributions at all times are fixed. In particular, we give the Duality Theorems for the SOTs and a sufficient condition for the finiteness of the minimum in the SOT with given initial and terminal distributions. We also show that the zero-noise limit of Schrödinger's problem exists and the limit solves Monge's problem with a quadratic cost and that the zero-noise limit of the Duality Theorem for the SOT gives the Duality Theorem for the OT.

2.1 Optimal Transportation Problem

In this section, we briefly discuss the OT by Kantorovich's approach on the state space $\mathbb{R}^d$ (see, e.g. [130, 152, 153], and the references therein).

$$\mathscr{A}(P_0, P_1) := \{\mu \in \mathscr{P}(\mathbb{R}^d \times \mathbb{R}^d) | \mu_1 = P_0, \mu_2 = P_1\}, \tag{2.1}$$

where for $\mu \in \mathscr{P}(\mathbb{R}^d \times \mathbb{R}^d)$

$$\mu_1(dx) := \mu(dx \times \mathbb{R}^d), \quad \mu_2(dy) := \mu(\mathbb{R}^d \times dy). \tag{2.2}$$

We recall the OT from Chap. 1 (see (1.28)).

Definition 2.1 (Optimal Transportation Problem) Let $c(\cdot, \cdot) : \mathbb{R}^d \times \mathbb{R}^d \longrightarrow [0, \infty)$ be Borel measurable. For $P_0, P_1 \in \mathscr{P}(\mathbb{R}^d)$,

$$T(P_0, P_1) := \inf\left\{ \int_{\mathbb{R}^d \times \mathbb{R}^d} c(x, y)\mu(dxdy) \middle| \mu \in \mathscr{A}(P_0, P_1) \right\}. \tag{2.3}$$

Study when the minimizer of $T(P_0, P_1)$ exists uniquely and when the support of the minimizer is a subset of a graph of a mapping on $\mathbb{R}^d$.

T. Mikami, *Stochastic Optimal Transportation*, SpringerBriefs in Mathematics,
https://doi.org/10.1007/978-981-16-1754-6_2

Remark 2.1 (i) Since $P_0 \times P_1 \in \mathscr{A}(P_0, P_1)$, $\mathscr{A}(P_0, P_1) \neq \emptyset$. (ii) Suppose that $c(x, y) = |y - x|^p$ for $p \geq 1$. Then $W_p(P_0, P_1) := T(P_0, P_1)^{1/p}$ defines a metric, on $\mathscr{P}(\mathbb{R}^d)$, which is called the L^p-Wasserstein metric/distance. Besides, $T(P_n, P) \to 0$, $n \to \infty$ if and only if $\int_{\mathbb{R}^d} |x|^p P_n(dx) \to \int_{\mathbb{R}^d} |x|^p P(dx)$ and $P_n \to P$ weakly, $n \to \infty$. (iii) In the case where $c(x, y) = 1_{(0,\infty)}(|y - x|)$, by Strassen's Theorem or by the Duality Theorem (see Theorem 2.1), $T(P_0, P_1)$ is a total variation distance (see [152]).

The so-called Duality Theorem for $T(\cdot, \cdot)$ plays a crucial role in the study of the OT. Suppose that $c(\cdot, \cdot) \in C(\mathbb{R}^d \times \mathbb{R}^d; [0, \infty))$. For $f \in C(\mathbb{R}^d)$,

$$T^* f(x) := \sup\{f(y) - c(x, y) | y \in \mathbb{R}^d\}. \tag{2.4}$$

Since the mapping $P \mapsto T(P_0, P)$ is convex and is lower semicontinuous in a weak topology, and since the dual space of the space of all finite signed Borel measures on $\mathbb{R}^d$ with a weak topology is $C_b(\mathbb{R}^d)$ (see [48], Theorem 2.2.15 and Lemma 3.2.3), the following holds (see [67, 85, 130, 152]).

Theorem 2.1 (Duality Theorem) *Suppose that* $c(\cdot, \cdot) \in C(\mathbb{R}^d \times \mathbb{R}^d; [0, \infty))$. *Then for any* $P_0 \in \mathscr{P}(\mathbb{R}^d)$ *for which* $T(P_0, \cdot)$ *is not identically equal to infinity and for any* $P_1 \in \mathscr{P}(\mathbb{R}^d)$,

$$\begin{aligned} T(P_0, P_1) &= \sup\left\{ \int_{\mathbb{R}^d} f(x) P_1(dx) - \int_{\mathbb{R}^d} T^* f(x) P_0(dx) \middle| f \in C_b(\mathbb{R}^d) \right\} \\ &= \sup\left\{ \int_{\mathbb{R}^d} \varphi(1, x) P_1(dx) - \int_{\mathbb{R}^d} \varphi(0, x) P_0(dx) \middle| \right. \\ &\qquad \left. \varphi(1, y) - \varphi(0, x) \leq c(x, y), \varphi(t, \cdot) \in C_b(\mathbb{R}^d), t = 0, 1 \right\}. \end{aligned} \tag{2.5}$$

We briefly explain how Theorem 2.1 gives the solution of the OT. The set $\mathscr{A}(P_0, P_1)$ is compact in the weak topology since

$$\mu(A \times B) \leq \min(P_0(A), P_1(B)), \quad \mu \in \mathscr{A}(P_0, P_1), A, B \in \mathbf{B}(\mathbb{R}^d).$$

Since $\mu \mapsto \int_{\mathbb{R}^d \times \mathbb{R}^d} c(x, y) \mu(dxdy)$ is lower semicontinuous in the weak topology, one can easily prove the existence of a minimizer μ_o of $T(P_0, P_1)$, provided that $T(P_0, P_1)$ is finite. By the theory of C. Smith and M. Knott (see [145]) on the generalization of the notion of the subdifferential of a convex function, there exists a maximizer $\varphi(t, \cdot) \in L^1(P_t)$, $t = 0, 1$ in Theorem 2.1. In particular,

$$\varphi(1, y) = \varphi(0, x) + c(x, y) = \min_{z \in \mathbb{R}^d} (\varphi(0, z) + c(z, y)), \quad \mu_o\text{–a.s.}, \tag{2.6}$$

since

$$\varphi(1, y) - \varphi(0, x) \le c(x, y), \quad x, y \in \mathbb{R}^d,$$
$$\varphi(1, y) - \varphi(0, x) = c(x, y), \quad \mu_o\text{–a.s..}$$

If c is partially differentiable in x, then for $(x, y) \in supp(\mu_o)$ and $\tilde{x} \in supp(P_0)$, there exists $\theta \in (0, 1)$ such that

$$\varphi(0, x) - \varphi(0, \tilde{x}) \le c(\tilde{x}, y) - c(x, y) = \langle D_x c(x + \theta(\tilde{x} - x), y), \tilde{x} - x\rangle,$$

by the mean value theorem, where $\langle \cdot, \cdot \rangle$ denote the inner product in $\mathbb{R}^d$. In particular, if $D_x c$ is locally bounded, then $\varphi(0, x)$ is locally Lipschitz continuous and is differentiable dx–a.e. in $supp(P_0)$. If $c(x, y)$ and $\varphi(0, x)$ are differentiable in x, then

$$D_x\varphi(0, x) + D_x c(x, y) = 0 \quad \mu_o\text{–a.s..}$$

In addition, if $y \mapsto D_x c(x, y)$ is invertible for $x \in supp(P_0)$, then

$$y = D_x c(x, \cdot)^{-1}(-D_x\varphi(0, x)), \quad \mu_o - \text{a.s..}$$

We recall Monge's problem from Chap. 1 (see (1.27)): for $P_0, P_1 \in \mathcal{P}(\mathbb{R}^d)$,

$$T_M(P_0, P_1) := \inf\left\{ \int_{\mathbb{R}^d} c(x, \psi(x)) P_0(dx) \middle| P_0\psi^{-1} = P_1 \right\}. \tag{2.7}$$

We refer readers to [67] for the following.

Theorem 2.2 *Suppose that* $c(x, y) = L(y - x)$ *and that* $L : \mathbb{R}^d \longrightarrow [0, \infty)$ *is strictly convex, smooth,* $L(u) = L(v)$ *if* $|u| = |v|$ *and* $\frac{L(u)}{|u|} \to \infty, |u| \to \infty$. *Then for any* $P_0, P_1 \in \mathcal{P}(\mathbb{R}^d)$ *for which* $P_0(dx) \ll dx$ *and* $T(P_0, P_1)$ *is finite, there exists a locally Lipschitz continuous function* φ *such that* $P_0(dx)\delta_{x+DH(D\varphi(x))}(dy)$ *is the unique minimizer of* $T(P_0, P_1)$, *where*

$$H(z) := \sup\{\langle z, u\rangle - L(u) | u \in \mathbb{R}^d\}. \tag{2.8}$$

In particular, $x + DH(D\varphi(x))$ *is the unique minimizer of* $T_M(P_0, P_1)$.

Remark 2.2 (See [55]) Suppose that $c(x, y) = L(y - x)$ and that $L : \mathbb{R}^d \longrightarrow [0, \infty)$ is convex. For $f \in C(\mathbb{R}^d)$,

$$\varphi(t, x; f) := \begin{cases} \sup\limits_{y \in \mathbb{R}^d} \left(f(y) - (1 - t)L\left(\dfrac{y - x}{1 - t}\right)\right), & (t, x) \in [0, 1) \times \mathbb{R}^d, \\ f(x), & (t, x) \in \{1\} \times \mathbb{R}^d. \end{cases} \tag{2.9}$$

Then, for $(t, x) \in [0, 1] \times \mathbb{R}^d$,

$$\varphi(t, x; f) = \sup\left\{ f(x(1)) - \int_t^1 L(\dot{x}(s))ds \,\middle|\, x(\cdot) \in AC([t, 1]; \mathbb{R}^d), x(t) = x \right\}, \quad (2.10)$$

$$\varphi(0, x; f) = T^* f(x)$$

(see (2.4) for notation). Suppose that f is uniformly Lipschitz continuous and that $\frac{H(z)}{|z|} \to \infty$, $|z| \to \infty$. Then the following holds.

(i) If H is smooth, then $\varphi(t, x; f)$ is uniformly Lipschitz continuous and is the unique weak solution of the following Hamilton–Jacobi equation:

$$\partial_t \varphi(t, x) + H(D_x \varphi(t, x)) = 0, \quad (t, x) \in [0, 1) \times \mathbb{R}^d, \qquad (2.11)$$
$$\varphi(1, x) = f(x), \quad x \in \mathbb{R}^d,$$

where $\partial_t := \partial/\partial t$. (2.9) is called the Hopf–Lax formula for the solution of (2.11).

(ii) If f is bounded, then $\varphi(t, x; f)$ is bounded, uniformly Lipschitz continuous and is the unique viscosity solution of (2.11) (see, e.g. [42, 55, 89] for viscosity solutions).

Theorem 2.1 has the following version (see, e.g. [55, 152]).

Corollary 2.1 *Suppose that* $c(x, y) = L(y - x)$ *and that* $L : \mathbb{R}^d \longrightarrow [0, \infty)$ *is convex. Then for any* P_0 *and* $P_1 \in \mathcal{P}(\mathbb{R}^d)$,

$$T(P_0, P_1) = \sup\left\{ \int_{\mathbb{R}^d} \varphi(1, y) P_1(dy) - \int_{\mathbb{R}^d} \varphi(0, x) P_0(dx) \right\}, \qquad (2.12)$$

where the supremum is taken over all bounded continuous viscosity solutions of (2.11).

Historical Remark Since there are so many references on the OT, this remark might be incomplete. We would like to ask readers to find complete references by themselves. V. L. Levin proved the Duality Theorem in the case where the state space is compact (see [97]). It was generalized by H. G. Kellerer, using a functional analytic method, in a more general setting (see [85]). The proof by the minmax principle is given in [152] (see [20, 74] for the recent development). In the case where $c(x, y) \to \infty$, $|y - x| \to \infty$, the selection lemma that is useful in the stochastic optimal control theory gives a simple proof of the Duality Theorem (see [110]). In the case where $L = |u|^2$, $P_0(dx) \ll dx$, Y. Brenier (see [21, 22]) solved the OT (see [108] for a probabilistic alternative proof). W. Gangbo and R. J. McCann generalized Y. Brenier's result, including the case where $L = |u|^p$, $p > 1$ (see [67]). The discussion given below Theorem 2.1 is due to their paper. L. C. Evans and W. Gangbo solved the OT in the case where $L = |u|$ and P_0 and P_1

have mutually disjoint compact supports (see [56]). L. A. Caffarelli, M. Feldman, and R. J. McCann (see [26]) and N. S. Trudinger and Xu-Jia Wang (see [151]) independently proved that the minimizer of the OT in the case where $L = |u|^p$, $p > 1$ is convergent as $p \to 1$ and that the limit is a minimizer of the OT in the case where $L = |u|$. A minimizer of the OT is not unique in the case where $L = |u|$ (see [46]). If $d = 1$ and $F_0(x) := P_0((-\infty, x])$, $x \in \mathbb{R}$ is continuous, then $F_1^{-1}(F_0(x))$ and $F_1^{-1}(1 - F_0(x))$ are (unique if $p > 1$) minimizers of the Monge's problem in the case where $L = |u|^p$ and $L = -|u|^p$, $p \geq 1$, respectively (see [47, 124, 130]). In particular, in the case where $d = 1$, one can easily show that minimizers obtained in [26, 56, 151] are $P_0(dx)\delta_{F_1^{-1}(F_0(x))}(dy)$.

2.2 Stochastic Optimal Transportation Problems

In this section, we first state two kinds of the SOTs. One is with two fixed endpoint marginal distributions, and the other is with a fixed flow of one-time marginals. We give the Duality Theorems for them. In particular, we give the characterization of the minimizer of the SOT with two endpoint marginals by the FBSDE and construct h-path processes with two endpoint marginals by the SOT. We also give, by the Duality Theorem, a sufficient condition for the finiteness of the minimum in the SOT with two endpoint marginals. Schrödinger's problem is an important class of the SOT with two given endpoint marginals. We show that the zero-noise limit of Schrödinger's problem exists and that the limit solves Monge's problem with a quadratic cost. We also show that the zero-noise limit of the Duality Theorem for the SOT with two endpoint marginals gives the Duality Theorem for the OT. The result in this section includes our recent result [118].

We first state two kinds of the SOTs.

Let $\sigma(t, x) = (\sigma_{ij}(t, x))_{i,j=1}^d$, $(t, x) \in [0, 1] \times \mathbb{R}^d$ be a Borel measurable $d \times d$-matrix function. Let $\mathscr{A}$ denote the set of all $\mathbb{R}^d$-valued, continuous semimartingales $\{X(t)\}_{0 \leq t \leq 1}$ on a (possibly different) complete filtered probability space such that there exists a Borel measurable $\beta_X : [0, 1] \times C([0, 1]; \mathbb{R}^d) \longrightarrow \mathbb{R}^d$ for which

(i) $\omega \mapsto \beta_X(t, \omega)$ is $\mathbf{B}(C([0, t]; \mathbb{R}^d))_+$-measurable for all $t \in [0, 1]$,
(ii) $X(t) = X(0) + \int_0^t \beta_X(s, X)ds + \int_0^t \sigma(s, X(s))dW_X(s)$, $0 \leq t \leq 1$,
(iii)

$$E\left[\int_0^1 \left\{|\beta_X(t, X)| + |\sigma(t, X(t))|^2\right\} dt\right] < \infty.$$

Here $\mathbf{B}(C([0, t]; \mathbb{R}^d))$, $\mathbf{B}(C([0, t]; \mathbb{R}^d))_+$ and W_X denote the Borel σ-field of $C([0, t]; \mathbb{R}^d)$, $\cap_{s>t}\mathbf{B}(C([0, s]; \mathbb{R}^d))$, and a d-dimensional Brownian motion, respectively (see, e.g. [100]). $|\cdot| := \langle \cdot, \cdot \rangle^{1/2}$.

Remark 2.3 In the definition of $\mathscr{A}$, in the case where $\sigma(t,x)$ is nondegenerate, W_X can be taken to be an $(\mathscr{F}_t^X)$-Brownian motion, where $\mathscr{F}_t^X$ denotes $\sigma[X(s) : 0 \le s \le t]$. Otherwise (ii) means that $X(t) - X(0) - \int_0^t \beta_X(s, X)ds$ is a local martingale with a quadratic variational process $\int_0^t a(s, X(s))ds$, where

$$a(t,x) := \sigma(t,x)\sigma(t,x)^t$$

and $\sigma(t,x)^t$ denotes the transpose of $\sigma(t,x)$ (see, e.g. [75] on a representation theorem for local martingales).

Let $L : [0,1] \times \mathbb{R}^d \times \mathbb{R}^d \longrightarrow [0,\infty)$ be continuous and $\mathbb{R}^d \ni u \mapsto L(t,x;u)$ be convex for $(t,x) \in [0,1] \times \mathbb{R}^d$.

The first problem of the following can be considered the stochastic optimal control version of the OT (see [112, 113, 119], and the references therein). Schrödinger's problem (1.54)–(1.55) in Sect. 1.4 chooses an optimal X among $\mathscr{A}$ in the case where two endpoint marginals are fixed. The second problem of the following determines an optimal X among $\mathscr{A}$ in the case where marginals at all times are fixed. It is also the continuum limit of the first one (see [113] and the references therein).

Definition 2.2 (Stochastic Optimal Transportation Problems (See [113, 119]))

(1) For $P_0, P_1 \in \mathscr{P}(\mathbb{R}^d)$,

$$V(P_0, P_1) := \inf_{\substack{X \in \mathscr{A}, \\ P^{X(t)} = P_t, t=0,1}} E\left[\int_0^1 L(t, X(t); \beta_X(t, X))dt\right]. \tag{2.13}$$

(2) For $\{P_t\}_{0 \le t \le 1} \subset \mathscr{P}(\mathbb{R}^d)$,

$$\mathbf{V}(\{P_t\}_{0 \le t \le 1}) := \inf_{\substack{X \in \mathscr{A}, \\ P^{X(t)} = P_t, 0 \le t \le 1}} E\left[\int_0^1 L(t, X(t); \beta_X(t, X))dt\right]. \tag{2.14}$$

Remark 2.4 The sets of stochastic processes over which the infimums are taken in (2.13)–(2.14) can be empty. If P_1 is absolutely continuous with respect to the Lebesgue measure and σ satisfies a nice condition, then it is not empty in (2.13) (see [80]). On (2.14), see [28, 29, 31–35, 103, 111, 112, 123, 159]. Indeed, it is related to Nelson's problem which constructs a Markov process from the Fokker–Planck equation (see Sect. 3.2.1). See Lemma 2.1 in Sect. 2.2.2 for the reason why the set $\mathscr{A}$ is appropriate as the set over which the infimum is taken in our SOTs (2.13)–(2.14). The minimizer of $\mathbf{V}(\{P_t\}_{0 \le t \le 1})$ was considered in [103, 112] (see also [107]). If the set over which the infimum is taken is empty, then we set the infimum to infinity in any infimum of this monograph.

We introduce problems similar to the above for a flow of marginal distributions which satisfies the Fokker–Planck equation.

Let $b : [0, 1] \times \mathbb{R}^d \longrightarrow \mathbb{R}^d$ be measurable and $\{P_t\}_{0\le t\le 1} \subset \mathcal{P}(\mathbb{R}^d)$. By $(a, b) \in \mathbf{A}(\{P_t\}_{0\le t\le 1})$, we mean that $a, b \in L^1([0, 1] \times \mathbb{R}^d, dt P_t(dx))$ and the following holds: for any $f \in C_b^{1,2}([0, 1] \times \mathbb{R}^d)$ and $t \in [0, 1]$,

$$\int_{\mathbb{R}^d} f(t, x)P_t(dx) - \int_{\mathbb{R}^d} f(0, x)P_0(dx) \tag{2.15}$$
$$= \int_0^t ds \int_{\mathbb{R}^d} \left(\partial_s f(s, x) + \frac{1}{2}\langle a(s, x), D_x^2 f(s, x)\rangle + \langle b(s, x), D_x f(s, x)\rangle\right) P_s(dx).$$

Here $D_x^2 := (\partial^2/\partial x_i \partial x_j)_{i,j=1}^d$ and $\langle A, B\rangle := \sum_{i,j=1}^d A_{ij} B_{ij}$ for $A = (A_{ij})_{i,j=1}^d, B = (B_{ij})_{i,j=1}^d \in M(d, \mathbb{R})$. In the case where a is fixed, we write $b \in \mathbf{A}(\{P_t\}_{0\le t\le 1})$ for simplicity.

Remark 2.5 For $\{X(t)\}_{0\le t\le 1} \in \mathcal{A}$,

$$(b_X(t, x) := E[\beta_X(t, X)|(t, X(t) = x)])_{(t,x)\in[0,1]\times\mathbb{R}^d} \in \mathbf{A}(\{P^{X(t)}\}_{0\le t\le 1}). \tag{2.16}$$

Indeed, by Itô's formula (see, e.g. [75]), for any $f \in C_b^{1,2}([0, 1]\times\mathbb{R}^d)$ and $t \in [0, 1]$,

$$E[f(t, X(t))] - E[f(0, X(0))] \tag{2.17}$$
$$= \int_0^t E\Big[\partial_s f(s, X(s)) + \frac{1}{2}\langle a(s, X(s)), D_x^2 f(s, X(s))\rangle$$
$$+ \langle \beta_X(s, X), D_x f(s, X(s))\rangle\Big] ds$$
$$= \int_0^t E\Big[\partial_s f(s, X(s)) + \frac{1}{2}\langle a(s, X(s)), D_x^2 f(s, X(s))\rangle$$
$$+ \langle E[\beta_X(s, X)|(s, X(s))], D_x f(s, X(s))\rangle\Big] ds.$$

$$E\left[\int_0^1 |E[\beta_X(t, X)|(t, X(t))]|dt\right] \le E\left[\int_0^1 |\beta_X(t, X)|dt\right] < \infty. \tag{2.18}$$

The following can be considered versions of the SOTs for a flow of marginals that satisfy (2.15).

Definition 2.3 (SOTs for Marginal Flows (See [113]))

(1) For $P_0, P_1 \in \mathscr{P}(\mathbb{R}^d)$,

$$\mathrm{v}(P_0, P_1) := \inf_{\substack{b \in \mathbf{A}(\{Q_t\}_{0 \le t \le 1}),\\ Q_t = P_t, t=0,1}} \int_{[0,1]\times\mathbb{R}^d} L(t, x; b(t, x)) dt Q_t(dx). \tag{2.19}$$

(2) For $\{P_t\}_{0 \le t \le 1} \subset \mathscr{P}(\mathbb{R}^d)$,

$$\mathbf{v}(\{P_t\}_{0 \le t \le 1}) := \inf_{b \in \mathbf{A}(\{P_t\}_{0 \le t \le 1})} \int_{[0,1]\times\mathbb{R}^d} L(t, x; b(t, x)) dt P_t(dx). \tag{2.20}$$

We introduce relaxed versions of the problems above for marginal measures (see [70, 71], and references therein for related topics).

We write $\nu(dtdxdu) \in \tilde{\mathscr{A}}$ if the following holds. (i) $\nu \in \mathscr{P}([0,1] \times \mathbb{R}^d \times \mathbb{R}^d)$ and

$$\int_{[0,1]\times\mathbb{R}^d\times\mathbb{R}^d} (|a(t, x)| + |u|) \nu(dtdxdu) < \infty. \tag{2.21}$$

(ii) $\nu(dtdxdu) = dt\nu(t, dxdu)$, $\nu(t, dxdu) \in \mathscr{P}(\mathbb{R}^d \times \mathbb{R}^d)$, $\nu_1(t, dx), \nu_2(t, du) \in \mathscr{P}(\mathbb{R}^d)$, dt–a.e. (see (2.2) for notation) and $t \mapsto \nu_1(t, dx)$ has a weakly continuous version $\nu_{1,t}(dx) \in \mathscr{P}(\mathbb{R}^d)$ for which the following holds: for any $t \in [0, 1]$ and $f \in C_b^{1,2}([0, 1] \times \mathbb{R}^d)$,

$$\begin{aligned} &\int_{\mathbb{R}^d} f(t, x) \nu_{1,t}(dx) - \int_{\mathbb{R}^d} f(0, x) \nu_{1,0}(dx) \\ &= \int_{[0,t]\times\mathbb{R}^d\times\mathbb{R}^d} \mathscr{L}_{s,x,u} f(s, x) \nu(dsdxdu). \end{aligned} \tag{2.22}$$

Here

$$\mathscr{L}_{s,x,u} f(s, x) := \partial_s f(s, x) + \frac{1}{2}\langle a(s, x), D_x^2 f(s, x)\rangle + \langle u, D_x f(s, x)\rangle. \tag{2.23}$$

The following can be considered versions of the SOTs for marginal measures which satisfy (2.22).

Definition 2.4 (SOTs for Marginal Measures (See [113]))

(1) For $P_0, P_1 \in \mathscr{P}(\mathbb{R}^d)$,

$$\tilde{\mathrm{v}}(P_0, P_1) := \inf_{\substack{\nu \in \tilde{\mathscr{A}},\\ \nu_{1,t} = P_t, t=0,1}} \int_{[0,1]\times\mathbb{R}^d\times\mathbb{R}^d} L(t, x; u) \nu(dtdxdu). \tag{2.24}$$

(2) For $\{P_t\}_{0\le t\le 1} \subset \mathscr{P}(\mathbb{R}^d)$,

$$\tilde{\mathbf{v}}(\{P_t\}_{0\le t\le 1}) := \inf_{\substack{\nu\in\tilde{\mathscr{A}},\\ \nu_{1,t}=P_t, 0\le t\le 1}} \int_{[0,1]\times\mathbb{R}^d\times\mathbb{R}^d} L(t,x;u)\nu(dtdxdu). \tag{2.25}$$

Remark 2.6 If $b \in \mathbf{A}(\{P_t\}_{0\le t\le 1})$ and $X \in \mathscr{A}$, then $dtP_t(dx)\delta_{b(t,x)}(du) \in \tilde{\mathscr{A}}$ and $dtP^{(X(t),\beta_X(t,X))}(dxdu) \in \tilde{\mathscr{A}}$, respectively (see (2.15) and (2.17)). In particular, $dtP^{(X(t),\beta_X(t,X))}(dxdu)$ is the distribution of a $[0,1]\times\mathbb{R}^d\times\mathbb{R}^d$-valued random variable $(t, X(t), \beta_X(t,X))$. This is why we call (2.24)–(2.25) the SOTs for marginal measures (see also Lemma 2.2 given later). One can also identify $\{P_t\}_{0\le t\le 1} \subset \mathscr{P}(\mathbb{R}^d)$ with $dtP_t(dx) \in \mathscr{P}([0,1]\times\mathbb{R}^d)$ when $\mathbf{V}$, $\mathbf{v}$ and $\tilde{\mathbf{v}}$ are considered (see Theorem 2.6, Proposition 2.2, and also [112]).

We introduce assumptions and give the relations among and the properties of three classes of the optimization problems stated in Definitions 2.2–2.4 above:

(A.0.0). (i) $\sigma_{ij} \in C_b([0,1]\times\mathbb{R}^d)$, $i,j = 1,\cdots,d$. (ii) σ is nondegenerate.

(A.1). (i) $L \in C([0,1]\times\mathbb{R}^d\times\mathbb{R}^d; [0,\infty))$. (ii) For $(t,x) \in [0,1]\times\mathbb{R}^d$, $u \mapsto L(t,x;u)$ is convex.

(A.2).

$$\lim_{|u|\to\infty} \frac{\inf\{L(t,x;u) \mid (t,x) \in [0,1]\times\mathbb{R}^d\}}{|u|} = \infty.$$

The following implies that Definitions 2.2–2.4 are equivalent in our setting and why they are all called stochastic optimal transportation problems. It also plays a crucial role in the proof of the convexities and the lower semicontinuities of $V(P_0,P_1)$ and $\mathbf{V}(\{P_t\}_{0\le t\le 1})$.

Proposition 2.1 (See [118])

(i) Suppose that (A.1) holds. Then the following holds:

$$V(P_0,P_1) = \mathrm{v}(P_0,P_1) = \tilde{\mathrm{v}}(P_0,P_1), \quad P_0, P_1 \in \mathscr{P}(\mathbb{R}^d), \tag{2.26}$$

$$\mathbf{V}(\{P_t\}_{0\le t\le 1}) = \mathbf{v}(\{P_t\}_{0\le t\le 1}) = \tilde{\mathbf{v}}(\{P_t\}_{0\le t\le 1}), \quad \{P_t\}_{0\le t\le 1} \subset \mathscr{P}(\mathbb{R}^d). \tag{2.27}$$

(ii) Suppose, in addition, that (A.0.0,i) and (A.2) hold. Then there exist minimizers X of $V(P_0,P_1)$ and Y of $\mathbf{V}(\{P_t\}_{0\le t\le 1})$ for which

$$\beta_X(t,X) = b_X(t,X(t)), \quad \beta_Y(t,Y) = b_Y(t,Y(t)), \tag{2.28}$$

provided $V(P_0,P_1)$ and $\mathbf{V}(\{P_t\}_{0\le t\le 1})$ are finite, respectively.

(iii) *Suppose, in addition, that (A.0.0,ii) holds and that $u \mapsto L(t,x;u)$ is strictly convex for $(t,x) \in [0,1] \times \mathbb{R}^d$. Then for any minimizers X of $V(P_0,P_1)$ and Y of $\mathbf{V}(\{P_t\}_{0\le t\le 1})$, (2.28) holds and b_X and b_Y in (2.28) are unique on the support of $dtP^{X(t)}(dx)$ and $dtP^{Y(t)}(dx)$, provided $V(P_0,P_1)$ and $\mathbf{V}(\{P_t\}_{0\le t\le 1})$ are finite, respectively.*

The superposition principle is the construction of a semimartingale from the Fokker–Planck equation (2.15). It was originally proposed by E. Nelson in the case where the Fokker–Planck equation is satisfied by the square of the absolute value of the solution to Schrödinger's equation and was first proved by E. Carlen. We call the semimartingale the **Nelson process** (see [125, 126], and also [19, 28, 29, 31–35, 59, 103, 105, 111, 112, 123, 150, 159]). In the case where $a \equiv 0$, it was studied in [3, 106, 107]. In the case where the linear operator with the second-order differential operator and with the Lévy measure is considered, it was studied in [35, 133].

We describe Trevisan's superposition principle and prove Proposition 2.1.

Theorem 2.3 (Trevisan's Superposition Principle (See [150])) *Suppose that there exist $b : [0,1] \times \mathbb{R}^d \longrightarrow \mathbb{R}^d$ and $\{P_t\}_{0\le t\le 1} \subset \mathcal{P}(\mathbb{R}^d)$ such that $b \in \mathbf{A}(\{P_t\}_{0\le t\le 1})$. Then there exists a semimartingale $\{X(t)\}_{0\le t\le 1}$ for which the following holds:*

$$X(t) = X(0) + \int_0^t b(s, X(s))ds + \int_0^t \sigma(s, X(s))dW_X(s), \tag{2.29}$$

$$P^{X(t)} = P_t, \quad 0 \le t \le 1. \tag{2.30}$$

Proof of Proposition 2.1 We prove (i). For $\{X(t)\}_{0\le t\le 1} \in \mathcal{A}$, by Jensen's inequality,

$$E\left[\int_0^1 L(t, X(t); \beta_X(t, X))dt\right] \ge E\left[\int_0^1 L(t, X(t); b_X(t, X(t)))dt\right]. \tag{2.31}$$

Theorem 2.3 implies the first equalities of (2.26)–(2.27) (see Remark 2.5). For $\nu \in \tilde{\mathcal{A}}$,

$$b_\nu(t,x) := \int_{\mathbb{R}^d} u\nu(t,x,du), \tag{2.32}$$

where $\nu(t,x,du)$ denotes a regular conditional probability of ν given (t,x). Then by Jensen's inequality,

$$\int_{[0,1]\times\mathbb{R}^d\times\mathbb{R}^d} L(t,x;u)\nu(dtdxdu) \ge \int_0^1 dt \int_{\mathbb{R}^d} L(t,x;b_\nu(t,x))\nu_{1,t}(dx). \tag{2.33}$$

$b_\nu \in \mathbf{A}(\{\nu_{1,t}\}_{0\le t\le 1})$ from (2.22). Indeed, by Jensen's inequality,

$$\int_{[0,1]\times\mathbb{R}^d} |b_\nu(t,x)| dt \nu_{1,t}(dx) \le \int_{[0,1]\times\mathbb{R}^d\times\mathbb{R}^d} |u|\nu(dtdxdu) < \infty,$$

and for any $t \in [0,1]$ and $f \in C_b^{1,2}([0,1]\times\mathbb{R}^d)$,

$$\begin{aligned}&\int_{[0,t]\times\mathbb{R}^d\times\mathbb{R}^d} \langle u, D_x f(s,x)\rangle \nu(dsdxdu) \\ &= \int_0^t ds \int_{\mathbb{R}^d} \langle b_\nu(s,x), D_x f(s,x)\rangle \nu_{1,s}(dx).\end{aligned} \tag{2.34}$$

This implies the second equalities of (2.26)–(2.27).

We prove (ii). Minimizing sequences of $\tilde{v}$ and $\tilde{\mathbf{v}}$ are tight from Lemma 2.2 given later in Sect. 2.2.2. In particular, the limits are minimizers of $\tilde{v}$ and $\tilde{\mathbf{v}}$, respectively. (2.32)–(2.34) and Trevisan's superposition principle imply (ii).

We prove (iii). Equation (2.31) implies the first part. We prove the uniqueness of b_X and b_Y. For $b \in \mathbf{A}(\{P_t\}_{0\le t\le 1})$, $P_t(dx) \ll dx$, dt–a.e. from (A.0.0,ii), since $a, b \in L^1([0,1]\times\mathbb{R}^d, dtP_t(dx))$ (see [18, p. 1042, Corollary 2.2.2]). For $\{p_i(t,x)dx\}_{0\le t\le 1} \subset \mathscr{P}(\mathbb{R}^d)$, $b_i \in \mathbf{A}(\{p_i(t,x)dx\}_{0\le t\le 1})$, $i = 0, 1$ and $\lambda \in [0,1]$,

$$p_\lambda := (1-\lambda)p_0 + \lambda p_1, \quad b_\lambda := 1_{(0,\infty)}(p_\lambda)\frac{(1-\lambda)p_0 b_0 + \lambda p_1 b_1}{p_\lambda}. \tag{2.35}$$

Then $b_\lambda \in \mathbf{A}(\{p_\lambda(t,x)dx\}_{0\le t\le 1})$ and

$$\begin{aligned}&\int_0^1 dt \int_{\mathbb{R}^d} L(t,x;b_\lambda(t,x))p_\lambda(t,x)dx \\ &\le (1-\lambda)\int_0^1 dt \int_{\mathbb{R}^d} L(t,x;b_0(t,x))p_0(t,x)dx \\ &\quad +\lambda \int_0^1 dt \int_{\mathbb{R}^d} L(t,x;b_1(t,x))p_1(t,x)dx,\end{aligned} \tag{2.36}$$

where the equality holds if and only if

$$b_0(t,x) = b_1(t,x), \quad dtdx\text{–a.e. on } \{(t,x)\in[0,1]\times\mathbb{R}^d \mid p_0(t,x)p_1(t,x) > 0\},$$

which completes the proof. □

2.2.1 Duality Theorem and Its Applications

The Duality Theorems for V and $\mathbf{V}$, respectively, imply that V and $\mathbf{V}$ are the dual problems of a class of finite horizon stochastic optimal control problems (see Theorems 2.4 and 2.6 below). In this section, we describe the results in [112, 118, 119], and some generalizations.

In [119], σ was an identity matrix and we only considered the Duality Theorem for V and its application. In [112], we generalized it to the case where σ is not an identity (see also [20, 147] and Remark 2.7) and also considered the Duality Theorem for $\mathbf{V}$ and other additional problems stated later. Our recent result [118] by Proposition 2.1 gives simpler proofs of the Duality Theorems under a weaker assumption than [112, 119]. Besides, it also gives the Duality Theorems for v and $\mathbf{v}$, which was proved separately from V and $\mathbf{V}$ in [112].

We describe additional assumptions, recalling (A.1)–(A.2) given before Proposition 2.1. (A.0) below is stronger than (A.0.0,i) given after Definition 2.4.

(A.0). $\sigma_{ij} \in C_b^1([0,1] \times \mathbb{R}^d),\ i,j = 1,\cdots,d.$

(A.1). (i) $L \in C([0,1] \times \mathbb{R}^d \times \mathbb{R}^d; [0,\infty))$. (ii) For $(t,x) \in [0,1] \times \mathbb{R}^d$, $u \mapsto L(t,x;u)$ is convex.

(A.2).

$$\lim_{|u|\to\infty} \frac{\inf\{L(t,x;u) | (t,x) \in [0,1] \times \mathbb{R}^d\}}{|u|} = \infty.$$

(A.3). (i) $\partial L(t,x;u)/\partial t$ and $D_x L(t,x;u)$ are bounded on $[0,1] \times \mathbb{R}^d \times B_R$ for all $R > 0$, where $B_R := \{x \in \mathbb{R}^d | |x| \le R\}$. (ii) C_L is finite, where

$$C_L := \sup\left\{ \frac{L(t,x;u)}{1 + L(t,y;u)} \middle| 0 \le t \le 1, x, y, u \in \mathbb{R}^d \right\}.$$

$$H(t,x;z) := \sup\{\langle z,u\rangle - L(t,x;u) | u \in \mathbb{R}^d\}. \tag{2.37}$$

The following is a generalization of [112, 119] and can be proved by the modification of the idea in [112] due to Proposition 2.1.

Theorem 2.4 (Duality Theorem for V (See [118])) *Suppose that (A.0)–(A.3) hold. Then, for any $P_0, P_1 \in \mathscr{P}(\mathbb{R}^d)$,*

$$\begin{aligned} V(P_0,P_1) &= \mathrm{v}(P_0,P_1) = \tilde{\mathrm{v}}(P_0,P_1) \qquad (2.38)\\ &= \sup\left\{ \int_{\mathbb{R}^d} \varphi(1,x) P_1(dx) - \int_{\mathbb{R}^d} \varphi(0,x) P_0(dx) \right\}, \end{aligned}$$

where the supremum is taken over all bounded continuous viscosity solutions φ to the following HJB Eqn.: on $[0, 1) \times \mathbb{R}^d$,

$$\partial_t \varphi(t, x) + \frac{1}{2}\langle a(t, x), D_x^2 \varphi(t, x)\rangle + H(t, x; D_x \varphi(t, x)) = 0, \tag{2.39}$$

$$\varphi(1, \cdot) \in C_b^\infty(\mathbb{R}^d).$$

Remark 2.7 In our previous results [112, 119], we assumed the nondegeneracy of a and used the Cameron–Martin–Maruyama–Girsanov formula to prove the convexity of $P \mapsto V(P_0, P)$. We also required the following assumption to prove the lower semicontinuity of $P \mapsto V(P_0, P)$:

$$\Delta L(\varepsilon_1, \varepsilon_2) := \sup \frac{L(t, x; u) - L(s, y; u)}{1 + L(s, y; u)} \to 0 \quad \text{as } \varepsilon_1, \varepsilon_2 \downarrow 0,$$

where the supremum is taken over all (t, x) and $(s, y) \in [0, 1] \times \mathbb{R}^d$ for which $|t - s| \leq \varepsilon_1$, $|x - y| < \varepsilon_2$ and over all $u \in \mathbb{R}^d$. Our recent approach [118] by the superposition principle does not require these two conditions. In [147, Lemma 3.15] (see also section 3.2 in their paper), they used a property on the convex combination of probability measures on an enlarged space, which also allowed them not to assume the nondegeneracy of a. The approach to the lower-semicontinuity of $P \mapsto V(P_0, P)$ was the same as our previous results and they used the above condition. Though their setting is not the same as ours, their approach is interesting and might be generalized to the path space version of our recent approach (see [94] for the recent development).

We define the viscosity solution to (2.39).

Definition 2.5 (Viscosity Solution (See, e.g. [42, 89]))

Viscosity Subsolution $\varphi \in USC([0, 1] \times \mathbb{R}^d)$ is a viscosity subsolution of (2.39) if whenever $h \in C^{1,2}([0, 1) \times \mathbb{R}^d)$ and $\varphi - h$ takes its maximum at $(s, y) \in [0, 1) \times \mathbb{R}^d$,

$$\partial_s h(s, y) + \frac{1}{2}\langle a(s, x), D_y^2 h(s, y)\rangle + H(s, y; D_x h(s, y)) \geq 0. \tag{2.40}$$

Viscosity Supersolution $\varphi \in LSC([0, 1] \times \mathbb{R}^d)$ is a viscosity supersolution of (2.39) if whenever $h \in C^{1,2}([0, 1) \times \mathbb{R}^d)$ and $\varphi - h$ takes its minimum at $(s, y) \in [0, 1) \times \mathbb{R}^d$,

$$\partial_s h(s, y) + \frac{1}{2}\langle a(s, x), D_y^2 h(s, y)\rangle + H(s, y; D_x h(s, y)) \leq 0. \tag{2.41}$$

Viscosity Solution $\varphi \in C([0, 1] \times \mathbb{R}^d)$ is a viscosity solution of (2.39) if it is both a viscosity sub- and a supersolution of (2.39).

We introduce the following to replace φ in (2.38) by classical solutions to the HJB Eqn. (2.39).

(A.4). (i) "σ is an identity" or "σ is uniformly nondegenerate, $\sigma_{ij} \in C_b^{1,2}([0,1] \times \mathbb{R}^d)$, $i, j = 1, \cdots, d$ and there exist functions L_1 and L_2 so that $L = L_1(t,x) + L_2(t,u)$". (ii) $L(t,x;u) \in C^1([0,1] \times \mathbb{R}^d \times \mathbb{R}^d; [0,\infty))$ and is strictly convex in u. (iii) $L \in C_b^{1,2,0}([0,1] \times \mathbb{R}^d \times B_R)$ for any $R > 0$.

For any $n \geq 1$, $f \in UC_b(\mathbb{R}^d)$ and $(t,x) \in [0,1] \times \mathbb{R}^d$,

$$\varphi_n(t,x;f) := \sup_{\substack{X(t)=x, X \in \mathscr{A}_t, \\ |\beta_X(s,X)| \leq n}} E\left[f(X(1)) - \int_t^1 L(s,X(s);\beta_X(s,X))ds \right], \tag{2.42}$$

where we define $\mathscr{A}_t$ as $\mathscr{A}$ with $[0,1]$ replaced by $[t,1]$.

$$H_n(t,x;z) := \sup_{|u| \leq n} \{\langle z, u\rangle - L(t,x;u)\}.$$

Then (A.0), (A.1,i) and (A.3) imply that $\varphi_n(t,x;f)$ is the unique bounded continuous viscosity solution of the following HJB Eqn.: for $(t,x) \in [0,1) \times \mathbb{R}^d$,

$$\partial_t \varphi_n(t,x) + \frac{1}{2}\langle a(t,x), D_x^2 \varphi_n(t,x)\rangle + H_n(t,x; D_x\varphi_n(t,x)) = 0, \tag{2.43}$$
$$\varphi_n(1,x) = f(x)$$

(see [61, p. 188, Corollary 7.1, p. 223, Corollary 3.1 and p. 249, Theorem 9.1]). If (A.4) holds, then the HJB Eqn. (2.39) with $\varphi(1,\cdot) = f(\cdot) \in C_b^3(\mathbb{R}^d)$ has the unique classical solution $\varphi(t,x;f)$ in $C_b^{1,2}([0,1] \times \mathbb{R}^d)$. In particular,

$$\varphi(t,x;f) = \varphi_n(t,x;f), \quad (t,x) \in [0,1] \times \mathbb{R}^d,$$

provided

$$n \geq \sup\{|D_z H(t,x; D_x\varphi(t,x;f))| \,|\, (t,x) \in [0,1] \times \mathbb{R}^d\} \in [0,\infty).$$

This can be proved almost in the same way as in [61, p. 208, Lemma 11.3] (see Proposition 2.3 in Sect. 2.2.2 and also [61, pp. 169–170, Theorems 4.2 and 4.4]). Lemma 2.5 in Sect. 2.2.2 implies that the minimal bounded continuous viscosity solution of the HJB Eqn. (2.39) with $\varphi(1,\cdot) = f(\cdot) \in C_b^3(\mathbb{R}^d)$ is also the unique classical solution of it under (A.4). Since (A.4,i), (A.4,ii) and (A.4,iii) imply (A.0), (A.1) and (A.3,i), respectively, we easily have the following from Lemma 2.6. We omit the proof.

Corollary 2.2 (See [118]) *Suppose that (A.2), (A.3,ii) and (A.4) hold. Then (2.38) holds even if the supremum is taken over all classical solutions* $\varphi \in C_b^{1,2}([0,1]\times\mathbb{R}^d)$ *to the HJB Eqn. (2.39). Besides, for any* $P_0, P_1 \in \mathcal{P}(\mathbb{R}^d)$ *for which* $V(P_0, P_1)$ *is finite, a minimizer* $\{X(t)\}_{0\le t\le 1}$ *of* $V(P_0, P_1)$ *exists and the following holds: for any maximizing sequence* $\{\varphi_n\}_{n\ge 1}$ *of (2.38),*

$$0 = \lim_{n\to\infty} E\Bigg[\int_0^1 |L(t, X(t); \beta_X(t, X)) \tag{2.44}$$

$$-\{\langle \beta_X(t, X), D_x\varphi_n(t, X(t))\rangle - H(t, X(t); D_x\varphi_n(t, X(t)))\}|dt\Bigg].$$

In particular, there exists a subsequence $\{n_k\}_{k\ge 1}$ *for which*

$$\beta_X(t, X) = b_X(t, X(t)) = \lim_{k\to\infty} D_z H(t, X(t); D_x\varphi_{n_k}(t, X(t))), \quad dtdP\text{-a.e.}. \tag{2.45}$$

We introduce an additional assumption:

(A.5). $L(t, x; \cdot) \in C^2(\mathbb{R}^d)$ for $(t, x) \in [0, 1] \times \mathbb{R}^d$. $D_u^2 L(t, x; u)$ is bounded and uniformly nondegenerate on $[0, 1] \times \mathbb{R}^d \times \mathbb{R}^d$.

From Corollary 2.2, in the same way as in [119, Theorem 2.2], we can show that a minimizer of $V(P_0, P_1)$ satisfies a FBSDE.

Theorem 2.5 (See [119]) *Suppose that (A3,ii) and (A.4)–(A.5) hold. Then for any* $P_0, P_1 \in \mathcal{P}(\mathbb{R}^d)$ *for which* $V(P_0, P_1)$ *is finite, there exists the unique minimizer* $\{X(t)\}_{0\le t\le 1}$ *of* $V(P_0, P_1)$. *There also exist measurable functions* f_t *on* $\mathbb{R}^d$, $t = 0, 1$ *and a* $(\mathcal{F}_t^X)$*-continuous semimartingale* $\{Y(t)\}_{0\le t\le 1}$ *such that the following FBSDE is satisfied by* $\{(X(t), Y(t), Z(t) := D_u L(t, X(t); b_X(t, X(t)))\}_{0\le t\le 1}$*: for* $t \in [0, 1]$,

$$X(t) = X(0) + \int_0^t D_z H(s, X(s); Z(s))ds + \int_0^t \sigma(s, X(s))dW_X(s), \tag{2.46}$$

$$Y(t) = f_1(X(1)) - \int_t^1 L(s, X(s); D_z H(s, X(s); Z(s)))ds$$

$$- \int_t^1 \langle Z(s), \sigma(s, X(s))dW_X(s)\rangle$$

and that $Y(0) = f_0(X(0))$.

We consider h-path processes as an application of Corollary 2.2 and Theorem 2.5 and generalize [119, Corollary 2.3]. We state the following assumption:

(A.4)'. $\sigma(\cdot) = (\sigma_{ij}(\cdot))_{i,j=1}^d$ is uniformly nondegenerate, $\sigma_{ij} \in C_b^{1,2}([0,1] \times \mathbb{R}^d)$, $i, j = 1, \cdots, d$. There exist functions $\xi \in C_b^{1,2}([0,1] \times \mathbb{R}^d; \mathbb{R}^d)$ and $c \in C_b^{1,2}([0,1] \times \mathbb{R}^d)$ such that for $(t, x, u) \in [0,1] \times \mathbb{R}^d \times \mathbb{R}^d$,

$$L(t, x; u) = \frac{1}{2}\langle a(t,x)^{-1}(u - \xi(t,x)), u - \xi(t,x)\rangle + c(t,x). \tag{2.47}$$

Under (A.4)',

$$H(t, x; z) = \frac{1}{2}\langle a(t,x)z, z\rangle + \langle \xi(t,x), z\rangle - c(t,x), \quad (t,x,z) \in [0,1] \times \mathbb{R}^d \times \mathbb{R}^d. \tag{2.48}$$

Under (A.4)', for $f(x) \in C_b^3(\mathbb{R}^d)$, let $u(t, x; f)$ be the unique classical solution, in $C_b^{1,2}([0,1] \times \mathbb{R}^d)$, of the following linear second-order PDE:

$$\partial_t u(t,x) + \frac{1}{2}\langle a(t,x), D_x^2 u(t,x)\rangle + \langle \xi(t,x), D_x u(t,x)\rangle + c(t,x)u(t,x) = 0, \tag{2.49}$$
$$u(1,x) = \exp(f(x)),$$

$(t, x) \in [0, 1) \times \mathbb{R}^d$ (see, e.g. [66]). Then $\varphi(t, x; f) := \log u(t, x; f)$ is the unique classical solution, in $C_b^{1,2}([0,1] \times \mathbb{R}^d)$, of the HJB Eqn. (2.39) with $\varphi(1, \cdot) = f(\cdot)$. It is also the minimal bounded continuous viscosity solution of the HJB Eqn. (2.39) with $\varphi(1, \cdot) = f(\cdot)$ since (A.4)' implies (A.0)–(A.3) (see Lemma 2.5 in Sect. 2.2.2). Indeed, by Itô's formula, $\varphi(t, x; f)$ has the same representation formula as that of the minimal bounded continuous viscosity solution of the HJB Eqn. (2.39).

Let $\{\mathbf{X}(t)\}_{0 \le t \le 1}$ be the unique strong solution to the following SDE (see [75]):

$$\mathbf{X}(t) = \mathbf{X}(0) + \int_0^t \xi(s, \mathbf{X}(s))ds + \int_0^t \sigma(s, \mathbf{X}(s))dW(s), \quad t \in [0,1]. \tag{2.50}$$

The following holds even though (A.4)' does not imply (A.4).

Corollary 2.3 (See [119]) *Suppose that (A.4)' holds. Then the assertions in Corollary 2.2 and Theorem 2.5 hold. Besides, for any $P_0, P_1 \in \mathcal{P}(\mathbb{R}^d)$ for which $V(P_0, P_1)$ is finite and for the unique minimizer $\{X(t)\}_{0 \le t \le 1}$ of $V(P_0, P_1)$, the following holds: for any $A \in \mathbf{B}(C([0,1]; \mathbb{R}^d))$,*

$$P(X(\cdot) \in A) = E\left[\exp\left\{f_1(\mathbf{X}(1)) - f_0(\mathbf{X}(0)) - \int_0^1 c(t, \mathbf{X}(t))dt\right\}; \mathbf{X}(\cdot) \in A\right]. \tag{2.51}$$

Remark 2.8 If (2.47) holds, then

$$L(t,x;u)-\{\langle u,z\rangle-H(t,x;z)\}=\frac{1}{2}|\sigma(t,x)^{-1}(a(t,x)z-u+\xi(t,x))|^2. \quad (2.52)$$

$\{X(t)\}_{0\le t\le 1}$ in Corollary 2.3 is called the h-path process for $\{\mathbf{X}(t)\}_{0\le t\le 1}$ with initial and terminal distributions P_0 and P_1, respectively. Equation (2.51) is known (see, e.g. [45, 62, 135, 156]).

We extend $\mathbf{V}$ to a variational problem on $\mathscr{P}([0,1]\times\mathbb{R}^d)$. For $\lambda\in\mathscr{P}([0,1]\times\mathbb{R}^d)$, we denote by $\lambda_t(dx)$ a weakly continuous version of $\lambda(t,dx):=\lambda(dtdx)/dt$, provided it exists.

Fix $P_0\in\mathscr{P}(\mathbb{R}^d)$. For $\lambda\in\mathscr{P}([0,1]\times\mathbb{R}^d)$,

$$\overline{\mathbf{V}}_{P_0}(\lambda):=\begin{cases}\mathbf{V}(\{\lambda_t\}_{0\le t\le 1}) & \text{if } \{\lambda_t\}_{0\le t\le 1} \text{ exists and } \lambda_0=P_0,\\ \infty, & \text{otherwise.}\end{cases} \quad (2.53)$$

Then it is easy to see that $\overline{\mathbf{V}}_{P_0}(\cdot)$ is convex, lower semi-continuous and not identically equal to infinity on $\mathscr{P}([0,1]\times\mathbb{R}^d)$ under (A.0)–(A.2) and (A3,ii) (see Proposition 2.1 and Lemma 2.2 in Sect. 2.2.2). We only explain the convexity of $\lambda\mapsto\overline{\mathbf{V}}_{P_0}(\lambda)$. $\overline{\mathbf{V}}_{P_0}(\lambda)$ is infinite if $\{\lambda_t\}_{t\in[0,1]}$ are not marginal distributions of $X\in\mathscr{A}$. Suppose that $\overline{\mathbf{V}}_{P_0}(\lambda^i)$ is finite, $i=1,2$. For $\nu^i\in\tilde{\mathscr{A}}$, $i=1,2$,

$$r\nu^1+(1-r)\nu^2\in\tilde{\mathscr{A}},\quad (r\nu^1+(1-r)\nu^2)_{1,t}=r\nu^1_{1,t}+(1-r)\nu^2_{1,t},\quad 0\le t\le 1, r\in(0,1)$$

since (2.22) linear in ν. If $\nu^i_{1,t}=\lambda^i_t$, $0\le t\le 1$, $i=1,2$, then

$$\int_{[0,1]\times\mathbb{R}^d\times\mathbb{R}^d}L(t,x;u)(r\nu^1+(1-r)\nu^2)(dtdxdu)\ge\tilde{\mathbf{v}}(\{r\lambda^1_t+(1-r)\lambda^2_t\}_{0\le t\le 1}).$$

Taking the infimum on the left-hand side over all $\nu^i\in\tilde{\mathscr{A}}$ such that $\nu^i_{1,t}=\lambda^i_t; 0\le t\le 1, i=1,2$,

$$r\tilde{\mathbf{v}}(\{\lambda^1_t\}_{0\le t\le 1})+(1-r)\tilde{\mathbf{v}}(\{\lambda^2_t\}_{0\le t\le 1})\ge\tilde{\mathbf{v}}(\{r\lambda^1_t+(1-r)\lambda^2_t\}_{0\le t\le 1}).$$

If $\lambda^i_0=P_0, i=1,2$, then $r\lambda^1_0+(1-r)\lambda^2_0=P_0$. Proposition 2.1 implies the convexity of $\lambda\mapsto\overline{\mathbf{V}}_{P_0}(\lambda)$.

For any $n\ge 1$, $f\in C_b^\infty([0,1]\times\mathbb{R}^d)$ and $(t,x)\in[0,1]\times\mathbb{R}^d$,

$$\phi_n(t,x;f) \quad (2.54)$$

$$:=\sup_{\substack{X(t)=x, X\in\mathscr{A}_t,\\ |\beta_X(s,X)|\le n}} E\left[\int_t^1(f(s,X(s))-L(s,X(s);\beta_X(s,X)))ds\right].$$

$$\phi(t,x;f):=\lim_{n\to\infty}\phi_n(t,x;f). \quad (2.55)$$

Here, notice that $n \mapsto \phi_n$ is nondecreasing. Then under (A.0)–(A.3), from Lemma 2.5 in Sect. 2.2.2, $\phi(t, x; f)$ is the minimal bounded continuous viscosity solution of the following HJB Eqn.: for $(t, x) \in [0, 1] \times \mathbb{R}^d$,

$$\partial_t \phi(t,x) + \frac{1}{2}\langle a(t,x), D_x^2\phi(t,x)\rangle + H(t,x; D_x\phi(t,x)) + f(t,x) = 0, \quad (2.56)$$

$$\phi(1,x) = 0.$$

In the same way as in Theorem 2.4 and Corollary 2.3, we obtain the Duality Theorem for $\mathbf{V}(\mathbf{P})$, the proof of which we omit.

Theorem 2.6 (Duality Theorem for V (See [118]))

(i) Suppose that (A.0)–(A.3) hold. Then for any $\mathbf{P} := \{P_t\}_{0\le t\le 1} \subset \mathscr{P}(\mathbb{R}^d)$,

$$\mathbf{V}(\mathbf{P}) = \mathbf{v}(\mathbf{P}) = \tilde{\mathbf{v}}(\mathbf{P}) \quad (2.57)$$

$$= \sup_{f\in C_b^\infty([0,1]\times\mathbb{R}^d)} \left\{ \int_0^1 \int_{\mathbb{R}^d} f(t,x)dtP_t(dx) - \int_{\mathbb{R}^d} \phi(0,x;f)P_0(dx) \right\}.$$

Suppose that (A.4) holds instead of (A.0), (A.1), and (A.3,i). Then (2.57) holds even if the supremum is taken over all classical solutions $\phi \in C_b^{1,2}([0,1]\times\mathbb{R}^d)$ *to the HJB Eqn. (2.56). Besides, if* $\mathbf{V}(\mathbf{P})$ *is finite, then a minimizer* $\{X(t)\}_{0\le t\le 1}$ *of* $\mathbf{V}(\mathbf{P})$ *exists and the following holds: for any maximizing sequence* $\{\phi_n\}_{n\ge 1}$ *of (2.57),*

$$0 = \lim_{n\to\infty} E\left[\int_0^1 |L(t, X(t); \beta_X(t,X)) \right. \quad (2.58)$$

$$\left. - \{\langle \beta_X(t,X), D_x\phi_n(t,X(t))\rangle - H(t,X(t); D_x\phi_n(t,X(t)))\}|dt \right].$$

In particular, there exists a subsequence $\{n_k\}_{k\ge 1}$ *for which*

$$\beta_X(t,X) = b_X(t,X(t)) = \lim_{k\to\infty} D_z H(t,X(t); D_x\phi_{n_k}(t,X(t))), \quad dtdP\text{-a.e.}. \quad (2.59)$$

(ii) Suppose that (A.4)' holds. Then the assertion above also holds.

We discuss the relation between the Duality Theorem for $V(P_0, P_1)$ and that for $\mathbf{V}(\{P_t\}_{0\le t\le 1})$ for $\{P_t\}_{0\le t\le 1} \subset \mathscr{P}(\mathbb{R}^d)$. Suppose that (A.0)–(A.2) hold. Then for $P_0, P_1 \in \mathscr{P}(\mathbb{R}^d)$ for which $V(P_0, P_1)$ is finite and for a minimizer $X \in \mathscr{A}$ of $V(P_0, P_1)$ (see Proposition 2.1), the following holds:

$$V(P_0, P_1) = \mathbf{V}(\{P^{X(t)}\}_{0\le t\le 1}). \quad (2.60)$$

Indeed, for any $\{P_t\}_{0\le t\le 1} \subset \mathcal{P}(\mathbb{R}^d)$,

$$V(P_0, P_1) \le \mathbf{V}(\{P_t\}_{0\le t\le 1}),$$

where the equality holds if $\{P_t\}_{0\le t\le 1} = \{P^{X(t)}\}_{0\le t\le 1}$. In particular, roughly speaking, the Duality Theorem for $V(P_0, P_1)$ implies that for $\mathbf{V}(\{P^{X(t)}\}_{0\le t\le 1})$. We explain it. Let $\{\rho_n\}_{n\ge 1} \in C^\infty([0,1];[0,1])$ such that $\dot\rho_n(t) \le 0$ and that

$$\rho_n(t) = \begin{cases} 1, & 0 \le t \le 1 - \frac{1}{n}, \\ 0, & t = 1. \end{cases}$$

If φ is smooth and satisfies (2.39), then $\{\rho_n(t)\varphi(t,x)\}_{n\ge 1}$ satisfies (2.56) with

$$f = f_n(t,x) := -\dot\rho_n(t)\varphi(t,x) + \rho_n(t)H(t,x;D_x\varphi(t,x)) - H(t,x;\rho_n(t)D_x\varphi(t,x)).$$

$\rho_n(0)\varphi(0,x) = \varphi(0,x)$ and

$$\int_0^1 f_n(t,X(t))dt = \int_{1-\frac{1}{n}}^1 f_n(t,X(t))dt \to \varphi(1,X(1)), \quad n \to \infty. \tag{2.61}$$

We extend $\mathrm{v}(\cdot,\cdot)$ and $\mathbf{v}(\cdot)$ to variational problems on $\mathcal{P}([0,1]\times\mathbb{R}^d)$. For $\lambda \in \mathcal{P}([0,1]\times\mathbb{R}^d)$,

$$\begin{aligned} \overline{\mathrm{v}}(\lambda) &:= \begin{cases} \mathrm{v}(\lambda_0,\lambda_1) \text{ if } \lambda(dtdx) = \frac{1}{2}\delta_0(dt)\lambda_0(dx) + \frac{1}{2}\delta_1(dt)\lambda_1(dx), \\ \infty \quad \text{otherwise}, \end{cases} \\ \overline{\mathbf{v}}(\lambda) &:= \begin{cases} \mathbf{v}(\{\lambda_t\}_{0\le t\le 1}) \text{ if } \{\lambda_t\}_{0\le t\le 1} \text{ exists}, \\ \infty \quad \text{otherwise} \end{cases} \end{aligned} \tag{2.62}$$

(compare with (2.53)). For $f \in C_b([0,1]\times\mathbb{R}^d)$,

$$\begin{aligned} \overline{\mathrm{v}}^*(f) &:= \sup_{\lambda\in\mathcal{P}([0,1]\times\mathbb{R}^d)} \left\{ \int_{[0,1]\times\mathbb{R}^d} f(t,x)\lambda(dtdx) - \overline{\mathrm{v}}(\lambda) \right\}, \\ \overline{\mathbf{v}}^*(f) &:= \sup_{\lambda\in\mathcal{P}([0,1]\times\mathbb{R}^d)} \left\{ \int_{[0,1]\times\mathbb{R}^d} f(t,x)\lambda(dtdx) - \overline{\mathbf{v}}(\lambda) \right\}. \end{aligned} \tag{2.63}$$

For $\{\nu_n\}_{n\ge 1} \subset \tilde{\mathcal{A}}$, if $\left\{\int_{[0,1]\times\mathbb{R}^d\times\mathbb{R}^d} L(t,x;u)\nu_n(dtdxdu)\right\}_{n\ge 1}$ is bounded and if $\{\nu_n(dtdx\times\mathbb{R}^d)\}_{n\ge 1}$ is tight, then $\{(\nu_n)_{1,0}(dx)\}_{n\ge 1}$ is tight under (A.2) and the boundedness of a (see (2.78), (2.79), (2.81)). The following is known (see [112] for the proof).

Proposition 2.2 (See [112]) *Suppose that (A.0)–(A.2) hold and that $L(t, x; 0)$ is bounded. Then for any $\lambda \in \mathcal{P}([0,1] \times \mathbb{R}^d)$,*

$$\overline{v}(\lambda) = \sup_{f \in C_b([0,1]\times\mathbb{R}^d)} \left\{ \int_{[0,1]\times\mathbb{R}^d} f(t,x)\lambda(dtdx) - \overline{v}^*(f) \right\}, \tag{2.64}$$

$$\overline{\mathbf{v}}(\lambda) = \sup_{f \in C_b([0,1]\times\mathbb{R}^d)} \left\{ \int_{[0,1]\times\mathbb{R}^d} f(t,x)\lambda(dtdx) - \overline{\mathbf{v}}^*(f) \right\}. \tag{2.65}$$

In particular, for any $\{P_t\}_{0\le t\le 1} \subset \mathcal{P}(\mathbb{R}^d)$, $V(P_0, P_1)$ and $\mathbf{V}(\{P_t\}_{0\le t\le 1})$ can be considered convex lower semicontinuous functionals of $\frac{1}{2}\delta_0(dt)P_0(dx) + \frac{1}{2}\delta_1(dt)P_1(dx)$ and $dtP_t(dx)$ in $\mathcal{P}([0,1]\times\mathbb{R}^d)$, respectively.

We discuss assumptions (A.0)–(A.5).

(A.0) implies the existence and the uniqueness of the solution to the following SDE:

$$dX(t) = \sigma(t, X(t))dW(t),$$

where W is a standard Brownian motion defined on a probability space.

(A.1,ii) implies that the infimums in $V(P_0, P_1)$ and $\mathbf{V}(\{P_t\}_{0\le t\le 1})$ can be taken for all X such that $\beta_X(t, X) = b_X(t, X(t))$ (see (2.31)).

(A.0), (A.1,i), and (A.2) imply the compactness result given in Lemma 2.2. In particular, (A.0)–(A.2) implies the duality (2.101). Indeed, we originally assumed that the diffusion matrix a is nondegenerate to prove the convexity of V and $\mathbf{V}$ by the change of measure. It can be proved by Proposition 2.1, without the nondegeneracy of a, in which the superposition principle plays a crucial role. This is a recent remarkable progress thanks to the superposition principle.

(A.3) is required to formulate Theorem 2.4 in the framework of viscosity solution. If $A(t,x) \in C_b^1([0,1]\times\mathbb{R}^d)$, $\inf\{A(t,x) | (t,x) \in [0,1]\times\mathbb{R}^d\}$ is positive, and $C(t,x) \in C_b^1([0,1]\times\mathbb{R}^d; [0,\infty))$, then $L = A(t,x)|u|^p + C(t,x)$ satisfies (A.1) and (A.3) if $p \ge 1$ and (A.2) if $p > 1$, respectively.

(A.4) is more restrictive than (A.3) and is required to formulate the Duality Theorem in the framework of a classical solution. In particular, a is required to be uniformly nondegenerate. $L = A(t,x)|u|^p + C(t,x)$ satisfies (A.4, ii, iii) if $A \in C_b^{1,2}([0,1]\times\mathbb{R}^d; (0,\infty))$, $C \in C_b^{1,2}([0,1]\times\mathbb{R}^d; [0,\infty))$ and $p > 1$.

(A.3)–(A.4) might be weakened by the PDE experts.

If $\tilde{A} \in C_b([0,1]\times\mathbb{R}^d; M_d(\mathbb{R}))$ and is uniformly nondegenerate, $B \in C([0,1]\times\mathbb{R}^d; \mathbb{R}^d)$, and $C \in C([0,1]\times\mathbb{R}^d)$, then $L = \langle \tilde{A}(t,x)u, u\rangle + \langle B(t,x), u\rangle + C(t,x)$ satisfies (A.5).

2.2.2 *Proof of Duality Theorem and Its Application*

In this section, we prove Theorems 2.4 and 2.5 and Corollary 2.3.

We first give technical lemmas and then prove Theorem 2.4.

Lemma 2.1 *Suppose that (A.0) holds. Let $(\Omega, \mathscr{F}, \{\mathscr{F}_t\}_{t\geq 0}, P)$, X_0 and $\{W(t)\}_{t\geq 0}$ be a complete filtered probability space, an $(\mathscr{F}_0)$-adapted random variable, and a d-dimensional $(\mathscr{F}_t)$-Brownian motion for which $W(0)=0$, respectively. For an $\mathbb{R}^d$-valued, $(\mathscr{F}_t)$-progressively measurable stochastic process $\{u(t)\}_{0\leq t\leq 1}$, consider a solution to the following:*

$$X^u(t) = X_0 + \int_0^t u(s)ds + \int_0^t \sigma(s, X^u(s))dW(s), \quad t \in [0,1]. \tag{2.66}$$

If $E[\int_0^1 |u(t)|dt]$ is finite, then $\{X^u(t)\}_{0\leq t\leq 1} \in \mathscr{A}$ with $\beta_{X^u}(t, X^u) = E[u(t)|\mathscr{F}_t^{X^u}]$, $dtdP$–a.e.. In particular, by Jensen's inequality,

$$E\left[\int_0^1 L(t, X^u(t); u(t))dt\right] \geq E\left[\int_0^1 L(t, X^u(t); \beta_{X^u}(t, X^u))dt\right]. \tag{2.67}$$

Proof of Proposition 2.1 Equation (2.66) has a unique strong solution from (A.0). We give a sketch of the proof. We first prove the existence.

$$X_0(t) := X_0 + \int_0^t u(s)ds, \tag{2.68}$$

$$X_{n+1}(t) := X_0 + \int_0^t u(s)ds + \int_0^t \sigma(s, X_n(s))dW(s), \quad t \in [0,1], n \geq 0.$$

Then

$$E\left[\sum_{n=0}^{\infty} \sup_{0\leq t\leq 1} |X_{n+1}(t) - X_n(t)|\right] < \infty. \tag{2.69}$$

Indeed,

$$X_{n+1}(t) - X_n(t) = \begin{cases} \int_0^t \{\sigma(s, X_n(s)) - \sigma(s, X_{n-1}(s))\}dW(s), & t \in [0,1], n \geq 1, \\ \int_0^t \sigma(s, X_0(s))dW(s), & t \in [0,1], n = 0. \end{cases}$$

Let L_σ denote the Lipschitz constant of σ. Then, by Doob's inequality and by induction, for $n \geq 1$,

$$E\left[\sup_{0\le s\le t} |X_{n+1}(s) - X_n(s)|^2\right] \le 4E\left[\int_0^t |\sigma(s, X_n(s)) - \sigma(s, X_{n-1}(s))|^2 ds\right]$$
$$\le 4L_\sigma^2 \int_0^t E\left[\sup_{0\le \alpha\le s} |X_n(\alpha) - X_{n-1}(\alpha)|^2\right] ds$$
$$\le (4L_\sigma^2)^n \frac{t^{n+1}}{(n+1)!}||\sigma||_\infty^2, \quad t \in [0, 1].$$

Notice that $X_{n+1}(s) - X_n(s)$ is square integrable from (A.0) though $X_n(s)$ is not necessarily.

Equation (2.69) implies that $\{\{X_n(t)\}_{0\le t\le 1}; n \geq 1\}$ is a Cauchy sequence of $C([0, 1]; \mathbb{R}^d)$, a.s. In particular, there exists $\{X_\infty(t)\}_{0\le t\le 1}$ such that the following holds:

$$\lim_{n\to\infty} \sup_{0\le t\le 1} |X_n(t) - X_\infty(t)| = 0, \quad \text{a.s.}, \tag{2.70}$$

$$E\left[\sup_{0\le t\le 1}\left|\int_0^t \{\sigma(s, X_n(s)) - \sigma(s, X_\infty(s))\}dW(s)\right|^2\right] \tag{2.71}$$
$$\le 4E\left[\int_0^1 |\sigma(s, X_n(s)) - \sigma(s, X_\infty(s))|^2 ds\right] \to 0, \quad n \to \infty$$

by Doob's inequality and by the bounded convergence theorem.

Taking a subsequence if necessary, let $n \to \infty$ in (2.68). Then from (2.70) and (2.71), one can show that $\{X_\infty(t)\}_{0\le t\le 1}$ is a solution to (2.66).

We prove the uniqueness. If (2.66) holds with $u = u_i$, $i = 1, 2$, then

$$X^{u_1}(t) - X^{u_2}(t) = \int_0^t \{\sigma(s, X^{u_1}(s)) - \sigma(s, X^{u_2}(s))\}dW(s), \quad t \in [0, 1].$$

In the same way as above, by Doob's inequality, for $t \in [0, 1]$,

$$E\left[\sup_{0\le s\le t} |X^{u_1}(s) - X^{u_2}(s))|^2\right] \le 4L_\sigma^2 \int_0^t E\left[\sup_{0\le \alpha\le s} |X^{u_1}(\alpha) - X^{u_2}(\alpha)|^2\right] ds.$$

Gronwall's inequality implies that $X^{u_1} = X^{u_2}$.

Since $E[\int_0^1 |u(t)|dt]$ is finite, there exists a Borel measurable $\beta_{X^u} : [0,1] \times C([0,1]) \longrightarrow \mathbb{R}^d$ for which $\omega \mapsto \beta_{X^u}(t,\omega)$ is $\mathbf{B}(C([0,t]))_+$-measurable for all $t \in [0,1]$ and for which

$$E[u(t)|\mathscr{F}_t^{X^u}] = \beta_{X^u}(t, X^u) \tag{2.72}$$

(see [100, pp. 114 and 270]). To complete the proof, we prove that

$$Y(t) := X^u(t) - X_0 - \int_0^t \beta_{X^u}(s, X^u)ds$$

is an $(\mathscr{F}_t^{X^u})$-martingale with a quadratic variational processes $(\int_0^t a_{ij}(s, X^u(s)) ds)_{i,j=1}^d$. Indeed, if this is true, then the martingale representation theorem (see, e.g. [75]) implies that

$$X^u(t) = X_0 + \int_0^t \beta_{X^u}(s, X^u)ds + \int_0^t \sigma(s, X^u(s))dW_{X^u}(s), \quad 0 \le t \le 1. \tag{2.73}$$

It is easy to see that $Y(t)$ is $\mathscr{F}_t^{X^u}$-adapted. Since $\mathscr{F}_s^{X^u} \subset \mathscr{F}_s$, if $0 \le s \le t$, then

$$\begin{aligned}
&E[Y(t) - Y(s)|\mathscr{F}_s^{X^u}] \qquad (2.74)\\
&= E\left[\int_s^t (u(\gamma) - \beta_{X^u}(\gamma, X^u))d\gamma + \int_s^t \sigma(\gamma, X^u(\gamma))dW(\gamma)\middle|\mathscr{F}_s^{X^u}\right]\\
&= \int_s^t E[u(\gamma) - E[u(\gamma)|\mathscr{F}_\gamma^{X^u}]|\mathscr{F}_s^{X^u}]d\gamma\\
&\quad + E\left[E\left[\int_s^t \sigma(\gamma, X^u(\gamma))dW(\gamma)\middle|\mathscr{F}_s\right]\middle|\mathscr{F}_s^{X^u}\right] = 0.
\end{aligned}$$

For any $f \in C_b^2(\mathbb{R})$ and $i, j = 1, \cdots, d$, by the Itô formula (see, e.g. [75]),

$$\begin{aligned}
&E[f(Y_i(t) - Y_i(s))f(Y_j(t) - Y_j(s))|\mathscr{F}_s^{X^u}] \qquad (2.75)\\
&= E\left[\int_s^t \left\{a_{ij}(\gamma, X^u(\gamma))Df(Y_i(\gamma) - Y_i(s))Df(Y_j(\gamma) - Y_j(s))\right.\right.\\
&\qquad + \frac{1}{2}a_{ii}(\gamma, X^u(\gamma))D^2 f(Y_i(\gamma) - Y_i(s))f(Y_j(\gamma) - Y_j(s))\\
&\qquad \left.\left. + \frac{1}{2}a_{jj}(\gamma, X^u(\gamma))f(Y_i(\gamma) - Y_i(s))D^2 f(Y_j(\gamma) - Y_j(s))\right\}d\gamma \middle| \mathscr{F}_s^{X^u}\right].
\end{aligned}$$

Indeed, since $Y(t)$ is $\mathscr{F}_t^{X^u}$-adapted, for $\gamma \geq s$,

$$\begin{aligned}
&E[f(Y_i(\gamma) - Y_i(s))Df(Y_j(\gamma) - Y_j(s))(u_j(\gamma) - \beta_{X^u,j}(\gamma, X^u))|\mathscr{F}_s^{X^u}] \\
&= E[E[f(Y_i(\gamma) - Y_i(s))Df(Y_j(\gamma) - Y_j(s))(u_j(\gamma) - E[u_j(\gamma)|\mathscr{F}_\gamma^{X^u}])|\mathscr{F}_\gamma^{X^u}]|\mathscr{F}_s^{X^u}] \\
&= E[f(Y_i(\gamma) - Y_i(s))Df(Y_j(\gamma) - Y_j(s)) \\
&\qquad \times E[u_j(\gamma) - E[u_j(\gamma)|\mathscr{F}_\gamma^{X^u}]|\mathscr{F}_\gamma^{X^u}]|\mathscr{F}_s^{X^u}] = 0,
\end{aligned}$$

where $u(s) = (u_i(s))_{i=1}^d$, $\beta_{X^u}(s, X^u) = (\beta_{X^u,i}(s, X^u))_{i=1}^d$. For $n \geq 1$, take f such that $f(x) = x$ if $|x| \leq n$. Then on the set $\cap_{k=i,j}\{\sup_{s\leq\gamma\leq t} |Y_k(\gamma) - Y_k(s)| \leq n\}$,

$$E[(Y_i(t) - Y_i(s))(Y_j(t) - Y_j(s))|\mathscr{F}_s^{X^u}] = E\left[\int_s^t a_{ij}(\gamma, X^u(\gamma))d\gamma \middle| \mathscr{F}_s^{X^u}\right]. \tag{2.76}$$

Since $\sup_{s\leq\gamma\leq t} |Y(\gamma) - Y(s)|$ is finite a.s., the proof is complete. □

For any $s \geq 0$ and $P \in \mathscr{P}(\mathbb{R}^d)$,

$$\boldsymbol{\Psi}_P(s) := \left\{\nu \in \tilde{\mathscr{A}} \middle| \nu_{1,0} = P, \int_{[0,1]\times\mathbb{R}^d\times\mathbb{R}^d} L(t, x; u)\nu(dtdxdu) \leq s\right\}. \tag{2.77}$$

The following lemma plays a crucial role in this section (see [112]).

Lemma 2.2 *Suppose that (A.0.0,i), (A.1,i), and (A.2) hold. Then for any $s \geq 0$ and compact set $K \subset \mathscr{P}(\mathbb{R}^d)$, the set $\cup_{P\in K}\boldsymbol{\Psi}_P(s)$ is compact in $\mathscr{P}([0, 1] \times \mathbb{R}^d \times \mathbb{R}^d)$, where $\mathscr{P}(\mathbb{R}^d)$ and $\mathscr{P}([0, 1] \times \mathbb{R}^d \times \mathbb{R}^d)$ are endowed with a weak topology.*

Proof of Proposition 2.1 We only have to consider the case where $\cup_{P\in K}\boldsymbol{\Psi}_P(s) \neq \emptyset$. We first prove that $\cup_{P\in K}\boldsymbol{\Psi}_P(s)$ is tight. From (A.2), for sufficiently large $r > 0$,

$$\begin{aligned}
&\nu([0, 1] \times \mathbb{R}^d \times B_r^c) \qquad\qquad (2.78)\\
&\leq \frac{1}{r}\int_{[0,1]\times\mathbb{R}^d\times B_r^c} |u|\nu(dtdxdu) \\
&\leq \frac{1}{r}\int_{[0,1]\times\mathbb{R}^d\times\mathbb{R}^d} L(t, x; u)\nu(dtdxdu) \leq \frac{s}{r}, \quad \nu \in \cup_{P\in K}\boldsymbol{\Psi}_P(s).
\end{aligned}$$

Take $\psi \in C_b^\infty(\mathbb{R}^d; [0, 1])$ for which $\psi(x) = 0$ if $|x| \leq 1$ and $= 1$ if $|x| \geq 2$, and

$$\psi_r(x) := \psi\left(\frac{x}{r}\right).$$

Then, from (2.22), (A.0.0,i), and (A.2), there exists $C > 0$ such that for sufficiently large $r \geq 1$ and any $t \in [0, 1]$,

$$\nu_{1,t}(B_{2r}^c) \leq \int_{\mathbb{R}^d} \psi\left(\frac{x}{r}\right) \nu_{1,t}(dx) \tag{2.79}$$

$$= \int_{\mathbb{R}^d} \psi\left(\frac{x}{r}\right) \nu_{1,0}(dx)$$

$$+ \int_{[0,t]\times\mathbb{R}^d\times\mathbb{R}^d} \left(\frac{1}{2r^2}\left\langle a(s,x), D_x^2\psi\left(\frac{x}{r}\right)\right\rangle + \frac{1}{r}\left\langle u, D_x\psi\left(\frac{x}{r}\right)\right\rangle\right)\nu(dsdxdu)$$

$$\leq \nu_{1,0}(B_r^c) + \frac{C}{r}\left(1 + \int_{[0,1]\times\mathbb{R}^d\times\mathbb{R}^d} L(t,x;u)\nu(dtdxdu)\right)$$

$$\leq \nu_{1,0}(B_r^c) + \frac{C(1+s)}{r}, \quad \nu \in \cup_{P\in K}\Psi_P(s).$$

Since $\nu_{1,0} \in K$, (2.78)–(2.79) imply the tightness of $\cup_{P\in K}\Psi_P(s)$.

Next, we prove that $\cup_{P\in K}\Psi_P(s)$ is closed. Suppose that $\nu_n \in \cup_{P\in K}\Psi_P(s)$ and that $\nu_n \to \nu$ as $n \to \infty$ weakly. Then it is easy to see that

$$\int_{[0,1]\times\mathbb{R}^d\times\mathbb{R}^d} L(t,x;u)\nu(dtdxdu) \tag{2.80}$$

$$\leq \liminf_{n\to\infty} \int_{[0,1]\times\mathbb{R}^d\times\mathbb{R}^d} L(t,x;u)\nu_n(dtdxdu) \leq s$$

from (A1,i). We can also prove that $(\nu_n)_{1,0}$ is convergent. Indeed, for any $\tilde{f} \in C([0,1])$ and $f \in C_b^{1,2}([0,1]\times\mathbb{R}^d)$, multiply $\tilde{f}$ and the both sides of (2.22) with ν replaced by ν_n and then integrate in t. Then since $\nu_n \in \cup_{P\in K}\Psi_P(s)$, we have the following from (A.0.0,i) and (A.2):

$$\int_0^1 \tilde{f}(t)dt \int_{\mathbb{R}^d} f(0,x)(\nu_n)_{1,0}(dx) \tag{2.81}$$

$$= \int_{[0,1]\times\mathbb{R}^d\times\mathbb{R}^d} \left(\tilde{f}(t)f(t,x) - \left(\int_t^1 \tilde{f}(s)ds\right)\mathscr{L}_{t,x,u}f(t,x)\right)\nu_n(dtdxdu)$$

$$\to \int_{[0,1]\times\mathbb{R}^d\times\mathbb{R}^d} \left(\tilde{f}(t)f(t,x) - \left(\int_t^1 \tilde{f}(s)ds\right)\mathscr{L}_{t,x,u}f(t,x)\right)\nu(dtdxdu), \; n \to \infty,$$

$$= \int_0^1 \tilde{f}(t)dt\left(\int_{\mathbb{R}^d} f(t,x)\nu_1(t,dx) - \int_{[0,t]\times\mathbb{R}^d\times\mathbb{R}^d} \mathscr{L}_{s,x,u}f(s,x)\nu(dsdxdu)\right)$$

(see (2.22) for notation). Here we used the following: since $\nu_n(dt \times \mathbb{R}^d \times \mathbb{R}^d) = dt$,

$$\nu(dt \times \mathbb{R}^d \times \mathbb{R}^d) = dt. \tag{2.82}$$

Equation (2.81) implies that $\nu_{1,t}(dx)$ exists, ν satisfies (2.22) and $\nu_{1,0} = \lim_{n\to\infty}(\nu_n)_{1,0} \in K$ since K is compact. It is easy to see that (2.21) holds from (2.80) and that $\nu \in \tilde{\mathscr{A}}$. □

We recall the following result.

Lemma 2.3 (See (2.42) and [61, pp. 185–188]) *Suppose that (A.0) and (A.3) hold. Then for any $n \ge 1$, $f \in UC_b(\mathbb{R}^d)$, $t \in [0,1]$ and $Q \in \mathscr{P}(\mathbb{R}^d)$,*

$$\int_{\mathbb{R}^d} \varphi_n(t,x;f)Q(dx) = \sup\left\{ E\left[f(X(1)) - \int_t^1 L(s,X(s);\beta_X(s,X))ds \right] \middle| X \in \mathscr{A}_t, |\beta_X(s,X)| \le n, P^{X(t)} = Q \right\}. \tag{2.83}$$

From Lemma 2.1, (A.0) implies that for any $t \in [0,1]$, $X \in \mathscr{A}_t$, and $n \ge 1$, the following has a unique solution: for $s \in [t,1]$,

$$X_n(s) = X(t) + \int_t^s 1_{[0,n]}(|\beta_X(\alpha,X)|)\beta_X(\alpha,X)d\alpha + \int_t^s \sigma(\alpha, X_n(\alpha))dW_X(\alpha). \tag{2.84}$$

$X_n \in \mathscr{A}_t$ and the following holds (see (2.72)):

$$\beta_{X_n}(s,X_n) = E[1_{[0,n]}(|\beta_X(s,X)|)\beta_X(s,X)|\mathscr{F}_s^{X_n}].$$

The following also holds.

Lemma 2.4 *Suppose that (A.0) holds. Then for any $t \in [0,1]$ and $X \in \mathscr{A}_t$, there exists a subsequence $\{X_{n(k)}\}_{k\ge1} \subset \mathscr{A}_t$ of (2.84) such that*

$$\lim_{k\to\infty} \sup_{t\le s\le 1} |X_{n(k)}(s) - X(s)| = 0, \quad \text{a.s..} \tag{2.85}$$

Proof of Proposition 2.1 For the sake of simplicity, we assume that $t = 0$.

$$\tau_m := \inf\left\{ s \in [0,1] \middle| \int_0^s |\beta_X(\alpha,X)|d\alpha > m \right\} (\to \infty \text{ as } m \to \infty). \tag{2.86}$$

Then we have

$$\lim_{n\to\infty} E\left[\sup_{0\le s\le \tau_m} |X_n(s) - X(s)|^2 \right] = 0 \quad \text{for } m \ge 1. \tag{2.87}$$

This is true, since

$$X(s) - X_n(s)$$
$$= \int_0^s 1_{(n,\infty)}(|\beta_X(\alpha, X)|)\beta_X(\alpha, X)d\alpha + \int_0^s (\sigma(\alpha, X(\alpha)) - \sigma(\alpha, X_n(\alpha)))dW_X(\alpha).$$

By the Gronwall and Doob inequalities and a standard method, for $m \geq 1$,

$$E\Big[\sup_{0\leq s\leq \tau_m} |X_n(s) - X(s)|^2\Big]$$
$$\leq 2E\bigg[\bigg|\int_0^{\tau_m} 1_{(n,\infty)}(|\beta_X(s, X)|)|\beta_X(s, X)|ds\bigg|^2\bigg]\exp(8L_\sigma^2) \to 0, \quad n \to \infty,$$

by the bounded convergence theorem. Here L_σ denotes the Lipschitz constant of σ and we used the following:

$$\int_0^{\tau_m} 1_{(n,\infty)}(|\beta_X(s, X)|)|\beta_X(s, X)|ds \leq m.$$

Since an L^2-convergent sequence of random variables has an a.s. convergent subsequence, one can take, by a diagonal method, a subsequence $\{n(k)\}_{k\geq 1}$ so that

$$\lim_{k\to\infty} \sup_{0\leq t\leq \tau_m} |X_{n(k)}(t) - X(t)| \to 0 \quad \text{for each } m \geq 1, \text{ a.s..} \tag{2.88}$$

Since $P(\cup_{m\geq 1}\{\tau_m = 1\}) = 1$, the proof is complete. □

$$\varphi(t, x; f) := \lim_{n\to\infty} \varphi_n(t, x; f)$$

(see (2.42) for notation). Then the following holds.

Lemma 2.5 *Suppose that (A.0)–(A.3) hold. Then for any* $f \in UC_b(\mathbb{R}^d)$, $Q \in \mathscr{P}(\mathbb{R}^d)$ *and* $t \in [0, 1]$,

$$\int_{\mathbb{R}^d} \varphi(t, x; f)Q(dx) \tag{2.89}$$
$$= \sup_{X\in\mathscr{A}_t, P^{X(t)}=Q} E\bigg[f(X(1)) - \int_t^1 L(s, X(s); \beta_X(s, X))ds\bigg].$$

$\varphi(t, x; f)$ is the minimal bounded continuous viscosity solution of (2.39) with $\varphi(1, x) = f(x)$.

Proof of Proposition 2.1 We write $\varphi(t,x;f)=\varphi(t,x)$ and $\varphi_n(t,x;f)=\varphi_n(t,x)$ for the sake of simplicity. We first prove (2.89). It is easy to see that the left-hand side is less than or equal to the right-hand side in (2.89) from Lemma 2.3. We prove the opposite inequality. For $X\in\mathscr{A}_t$ for which $P^{X(t)}=Q$ and for which $E[\int_t^1 L(s,X(s);\beta_X(s,X))ds]$ is finite, take $\{X_{n(k)}\}_{k\geq 1}$ in Lemma 2.4. Then for $k\geq 1$, from Lemma 2.3,

$$\begin{aligned}&\int_{\mathbb{R}^d}\varphi_{n(k)}(t,x)Q(dx) \qquad (2.90)\\ &\geq E\left[f(X_{n(k)}(1))-\int_t^1 L(s,X_{n(k)}(s);\beta_{X_{n(k)}}(s,X_{n(k)}))ds\right]\\ &\geq E\left[f(X_{n(k)}(1))-\int_t^1 L(s,X_{n(k)}(s);1_{[0,n(k)]}(|\beta_X(s,X)|)\beta_X(s,X))ds\right]\\ &\to E\left[f(X(1))-\int_t^1 L(s,X(s);\beta_X(s,X))ds\right]\quad \text{as } k\to\infty,\end{aligned}$$

by Jensen's inequality and the dominated convergence theorem, Indeed, (A.1,i) and (A.3,ii) imply that $L(t,x;0)$ is bounded and the following holds:

$$\begin{aligned}0&\leq L(s,X_{n(k)}(s);1_{[0,n(k)]}(|\beta_X(s,X)|)\beta_X(s,X)) \qquad (2.91)\\ &\leq (1+C_L)L(s,X(s);1_{[0,n(k)]}(|\beta_X(s,X)|)\beta_X(s,X))\\ &\leq (1+C_L)\{L(s,X(s);\beta_X(s,X))+L(s,X(s);0)\}.\end{aligned}$$

φ is bounded since setting $Q(dy)=\delta_x(dy)$ in (2.89), from (A.3,ii),

$$\begin{aligned}\sup_{y\in\mathbb{R}^d}f(y)\geq\varphi(t,x)&\geq E\left[f(X^0(1))-\int_t^1 L(s,X^0(s);0)ds\,\middle|\,X^0(t)=x\right] \qquad (2.92)\\ &\geq \inf_{y\in\mathbb{R}^d}f(y)-\sup_{(s,y)\in[0,1]\times\mathbb{R}^d}L(s,y;0)>-\infty\end{aligned}$$

(see (2.66) for notation).

We prove the upper semicontinuity of φ. For $(t,x)\in\mathbb{R}^d$ and $n\geq 1$, take $X_{n,t,x}\in\mathscr{A}_t$ for which $X_{n,t,x}(t)=x$ and

$$\varphi(t,x)-\frac{1}{n}<E\left[f(X_{n,t,x}(1))-\int_t^1 L(s,X_{n,t,x}(s);\beta_{X_{n,t,x}}(s,X_{n,t,x}))ds\right]. \qquad (2.93)$$

Then from (2.92),

$$E\left[\int_t^1 L(s, X_{n,t,x}(s); \beta_{X_{n,t,x}}(s, X_{n,t,x}))ds\right] \tag{2.94}$$
$$\leq 2 \sup_{y\in\mathbb{R}^d} |f(y)| + \sup_{(s,y)\in[0,1]\times\mathbb{R}^d} L(s, y; 0) + 1 < \infty.$$

$X_{n,t,x}(u) := x$ for $u < t$.

$$\begin{aligned} &\nu_{n,t,x}(dsdydu) \\ &:= ds P^{(X_{n,t,x}(s), b_{X_{n,t,x}}(s, X_{n,t,x}(s)))}(dydu) \\ &= 1_{[0,t]}(s) ds \delta_x(dy)\delta_0(du) + 1_{(t,1]}(s) ds P^{X_{n,t,x}(s)}(dy)\delta_{b_{X_{n,t,x}}(s,y)}(du) \end{aligned} \tag{2.95}$$

(see (2.16) for notation). Then $\nu_{n,t,x} \in \tilde{\mathscr{A}}$ with $\mathscr{L}_{s,y,u}$ replaced by the following: for $f \in C^{1,2}([0, 1] \times \mathbb{R}^d)$,

$$\mathscr{L}_{t,s,y,u} f(s, y) := \partial_s f(s, y) + \frac{1}{2}\langle 1_{(t,1]}(s)a(s, y), D_x^2 f(s, y)\rangle + \langle u, D_x f(s, y)\rangle. \tag{2.96}$$

In the same way as in Lemma 2.2, one can show that as $(n, t, x) \to (\infty, t_0, x_0)$, $\nu_{n,t,x}$ has a convergent subsequence. Approximate $1_{(t_0,1]}$ by a bounded continuous function in $L^1([0, 1], dt)$. Then the limit $\nu_{t_0,x_0} \in \tilde{\mathscr{A}}$ with $\mathscr{L}_{s,y,u}$ replaced by $\mathscr{L}_{t_0,s,y,u}$ and the following holds:

$$1_{[0,t_0]}(s)\nu_{t_0,x_0}(dsdydu) = 1_{[0,t_0]}(s) ds \delta_{x_0}(dy)\delta_0(du).$$

In particular, by Jensen's inequality and Trevisan's superposition principle,

$$\begin{aligned} &\limsup_{(t,x)\to(t_0,x_0)} \varphi(t, x) \\ &\leq \int_{\mathbb{R}^d} f(y)(\nu_{(t_0,x_0)})_{1,1}(dy) - \int_{[t_0,1]\times\mathbb{R}^d\times\mathbb{R}^d} L(s, y; u)\nu_{(t_0,x_0)}(dsdydu) \\ &\leq \int_{\mathbb{R}^d} f(y)(\nu_{(t_0,x_0)})_{1,1}(dy) - \int_{[t_0,1]\times\mathbb{R}^d} L(s, y; b_{\nu_{(t_0,x_0)}}(s, y))ds(\nu_{(t_0,x_0)})_{1,s}(dy) \\ &\leq \varphi(t_0, x_0) \end{aligned} \tag{2.97}$$

(see (2.22) and (2.32) for notation).

We prove the lower semicontinuity of φ. Since $\varphi_n \uparrow \varphi, n \to \infty$ and $\varphi_n(t, x; f)$ is continuous,

$$\liminf_{(t,x)\to(t_0,x_0)} \varphi(t, x) \ge \liminf_{(t,x)\to(t_0,x_0)} \varphi_n(t, x) = \varphi_n(t_0, x_0) \to \varphi(t_0, x_0), \quad n \to \infty. \tag{2.98}$$

Since $\varphi_n(t, x; f)$ is the unique bounded continuous viscosity solution of the HJB Eqn. (2.39) with $\varphi_n(1, x; f) = f(x)$ and with H replaced by H_n, one can prove that φ is a viscosity solution of (2.39) by a standard argument. Indeed, $H_n \uparrow H$ (resp. $\varphi_n \uparrow \varphi$) uniformly on every compact subset of $[0, 1] \times \mathbb{R}^d \times \mathbb{R}^d$ (resp. $[0, 1] \times \mathbb{R}^d$) as $n \to \infty$ by Dini's Theorem, since $H_n \uparrow H$ (resp. $\varphi_n \uparrow \varphi$) as $n \to \infty$ and H_n and H (resp. φ_n and φ) are continuous.

Let $\overline{\varphi}$ be a bounded continuous viscosity solution of (2.39) with $\overline{\varphi}(1, x) = f(x)$. Then $\overline{\varphi}$ is a bounded continuous viscosity supersolution of (2.39) with $\overline{\varphi}(1, x) = f(x)$ and with H replaced by H_n for all $n \ge 1$, since $H_n \le H$. Since $\varphi_n(t, x; f)$ is a bounded continuous viscosity solution of (2.39) with $\varphi_n(1, x; f) = f(x)$ and with H replaced by H_n, by the comparison principle (see [61, p. 249, Theorem 9.1]), $\varphi_n(t, x; f) \le \overline{\varphi}(t, x)$. Letting $n \to \infty$, the proof is over. □

Remark 2.9 The right-hand side of (2.89) is a finite horizon stochastic optimal control problem (see [61]) and the following holds:

$$\sup_{X\in\mathcal{A}_t, PX(t)^{-1}=Q} E\left[f(X(1)) - \int_t^1 L(s, X(s); \beta_X(s, X))ds\right] \tag{2.99}$$

$$= \sup_{P\in\mathcal{P}(\mathbb{R}^d)} \left\{\int_{\mathbb{R}^d} f(x)P(dx) - V_{[t,1]}(Q, P)\right\}.$$

Here $V_{[t,1]}$ denotes V with $[0, 1]$ replaced by $[t, 1]$.

To prove Theorem 2.4, we improve the idea in [119, Theorem 2.1].

Proof of Theorem 2.4 $V(P_0, \cdot) \not\equiv \infty$. Indeed, for X^0 for which $P^{X^0(0)} = P_0$, $P_1 := P^{X^0(1)}$ (see (2.66) for notation). Then, from (A.1,i) and (A.3,ii),

$$V(P_0, P_1) \le \sup\{L(t, x; 0)|(t, x) \in [0, 1] \times \mathbb{R}^d\} < \infty. \tag{2.100}$$

Consider $P \mapsto V(P_0, P)$ as a function on the space of finite signed Borel measures on $\mathbb{R}^d$, by putting $V(P_0, P) = +\infty$ for $P \notin \mathcal{P}(\mathbb{R}^d)$. Then it is convex and lower semicontinuous in the weak topology from Proposition 2.1 and Lemma 2.2. From [48, Theorem 2.2.15 and Lemma 3.2.3],

$$V(P_0, P_1) = \sup_{f\in C_b(\mathbb{R}^d)} \left\{\int_{\mathbb{R}^d} f(x)P_1(dx) - V^*_{P_0}(f)\right\}, \tag{2.101}$$

where for $f \in C_b(\mathbb{R}^d)$,

$$V_{P_0}^*(f) := \sup_{P \in \mathscr{P}(\mathbb{R}^d)} \left\{ \int_{\mathbb{R}^d} f(x)P(dx) - V(P_0, P) \right\}. \tag{2.102}$$

Denote by $\mathscr{V}(P_0, P_1)$ the right-hand side of (2.38). Then, from Lemma 2.5 and (2.101),

$$V(P_0, P_1) \geq \mathscr{V}(P_0, P_1).$$

We prove the opposite inequality. Take $\rho \in C_0^\infty([-1, 1]^d)$ for which $\rho(x)dx \in \mathscr{P}(\mathbb{R}^d)$. For $\varepsilon > 0$ and $f \in C_b(\mathbb{R}^d)$,

$$\rho_\varepsilon(x) := \varepsilon^{-d}\rho(x/\varepsilon),$$
$$f_\varepsilon(x) := \int_{\mathbb{R}^d} f(y)\rho_\varepsilon(x - y)dy, \quad x \in \mathbb{R}^d.$$

Then $f_\varepsilon \in C_b^\infty(\mathbb{R}^d)$ and, from Lemma 2.5,

$$\mathscr{V}(P_0, P_1) \geq \int_{\mathbb{R}^d} f_\varepsilon(x)P_1(dx) - V_{P_0}^*(f_\varepsilon). \tag{2.103}$$

Take $X_\varepsilon \in \mathscr{A}$, $\varepsilon > 0$ for which $P^{X_\varepsilon(0)} = P_0$ and

$$V_{P_0}^*(f_\varepsilon) - \varepsilon < E[f_\varepsilon(X_\varepsilon(1))] - E\left[\int_0^1 L(t, X_\varepsilon(t); \beta_{X_\varepsilon}(t, X_\varepsilon))dt\right]. \tag{2.104}$$

Then from (2.100), $\{V_{P_0}^*(f_\varepsilon)\}_{\varepsilon>0}$ is bounded below and $dtP^{X_\varepsilon(t)}(dx)\delta_{b_{X_\varepsilon}(t,X_\varepsilon(t))}(du)$ has a convergent subsequence in $\tilde{\mathscr{A}}$ from Lemma 2.2 (see (2.16) for notation). By Trevisan's superposition principle, for any weak limit point ν, there exists $X_\nu \in \mathscr{A}$ such that

$$\beta_{X_\nu}(t, X_\nu) = b_\nu(t, X_\nu(t)), \quad dtdP\text{–a.e.}, \tag{2.105}$$
$$\nu(dtdx \times \mathbb{R}^d) = dtP^{X_\nu(t)}(dx)$$

(see (2.34) for notation). By Fatou's lemma and Jensen's inequality (see (2.32)–(2.33)),

$$\begin{aligned}\limsup_{\varepsilon\to 0} V_{P_0}^*(f_\varepsilon) &\leq E[f(X_\nu(1))] - \int_{[0,1]\times\mathbb{R}^d\times\mathbb{R}^d} L(t, x; u)\nu(dtdxdu) \quad (2.106)\\ &\leq E[f(X_\nu(1))] - E\left[\int_0^1 L(t, X_\nu(t); b_\nu(t, X_\nu(t)))dt\right]\\ &\leq V_{P_0}^*(f)\end{aligned}$$

since

$$E[f_\varepsilon(X_\varepsilon(1))] = \int_{\mathbb{R}^d} \rho(z)dzE[f(X_\varepsilon(1)+\varepsilon z)].$$

Equations (2.101), (2.103), and (2.106) imply that

$$\mathscr{V}(P_0, P_1) \geq V(P_0, P_1),$$

which completes the proof. □

We give the following for the sake of completeness.

Proposition 2.3 *Suppose that (A.2) and (A.4,ii,iii) hold. Then, for any* $r > 0$,

$$\sup\{|D_z H(t,x;z)| | (t,x,z) \in [0,1] \times \mathbb{R}^d \times \mathbb{R}^d, |z| < r\} < \infty. \tag{2.107}$$

Proof of Theorem 2.4 For any $r > 0$, there exists $R(r) > 0$ such that

$$\inf\{|D_u L(t,x;u)| | (t,x,u) \in [0,1] \times \mathbb{R}^d \times \mathbb{R}^d, |u| > R(r)\} \geq r. \tag{2.108}$$

Indeed,

$$\begin{aligned}
&\inf\{|D_u L(t,x;u)| | (t,x) \in [0,1] \times \mathbb{R}^d\} \\
&\geq \frac{1}{|u|} \inf\{L(t,x;u) - L(t,x;0) | (t,x) \in [0,1] \times \mathbb{R}^d\} \to \infty, \quad |u| \to \infty,
\end{aligned}$$

from (A.2) and (A4,ii,iii) since from (A.4,ii), for any $(t,x,u) \in [0,1] \times \mathbb{R}^d \times \mathbb{R}^d$,

$$L(t,x;0) \geq L(t,x;u) + \langle D_u L(t,x;u), -u \rangle.$$

The supremum in (2.107) is less than or equal to $R(r)(< \infty)$. Indeed, if this is not true, then there exists $(t,x,z) \in [0,1] \times \mathbb{R}^d \times \mathbb{R}^d$ for which

$$|z| < r, \quad |D_z H(t,x;z)| > R(r).$$

The second inequality implies that the following holds: from (2.108),

$$|z| = |D_u L(t,x;D_z H(t,x;z))| \geq r$$

from (A.2) and (A.4,ii) (see, e.g. [152, 2.1.3]), which is a contradiction. □

As a preparation for the proof of Theorem 2.5, we give the following lemma which also implies (2.45).

Lemma 2.6 *Suppose that (A.2), (A.3,ii), and (A.4) hold. Then, for any maximizing sequence* $\{\varphi_n\}_{n\geq 1}$ *of Corollary 2.2 and any minimizer* $\{X(t)\}_{0\leq t\leq 1}$ *of* $V(P_0, P_1)$, $P_0, P_1 \in \mathscr{P}(\mathbb{R}^d)$, *which is finite, (2.44)–(2.45) hold.*

Proof of Theorem 2.4 The duality in Corollary 2.2 implies, by Itô's formula, that the following holds:

$$E\left[\int_0^1 L(t, X(t); \beta_X(t, X))dt\right] \tag{2.109}$$

$$= \lim_{n\to\infty} E\left[\int_0^1 \{\langle \beta_X(t, X), D_x\varphi_n(t, X(t))\rangle - H(t, X(t); D_x\varphi_n(t, X(t)))\}dt\right].$$

From (A.1), the following holds (see, e.g. [152]): for any $(t, x, u) \in [0, 1]\times\mathbb{R}^d\times\mathbb{R}^d$,

$$L(t, x; u) = \sup_{z\in\mathbb{R}^d}\{\langle z, u\rangle - H(t, x; z)\}. \tag{2.110}$$

Therefore (2.109) implies (2.44). Equation (2.45) is true from (2.44), (A.2), and (A.4,ii) (see, e.g. [152, 2.1.3]). □

Proof of Theorem 2.5 Take a maximizing sequence $\{\varphi_n\}_{n\geq 1}$ of Corollary 2.2. Then for any minimizer $\{X(t)\}_{0\leq t\leq 1}$ of $V(P_0, P_1)$, by Itô's formula,

$$\varphi_n(t, X(t)) - \varphi_n(0, X(0)) \tag{2.111}$$

$$= \int_0^t \{\langle b_X(s, X(s)), D_x\varphi_n(s, X(s))\rangle - H(s, X(s); D_x\varphi_n(s, X(s)))\}ds$$

$$+ \int_0^t \langle D_x\varphi_n(s, X(s)), \sigma(s, X(s))dW_X(s)\rangle, \quad t \in [0, 1].$$

By Doob's inequality (see [75]),

$$E\left[\sup_{0\leq t\leq 1}\left|\int_0^t \langle D_x\varphi_n(s, X(s)), \sigma(s, X(s))dW_X(s)\rangle\right.\right. \tag{2.112}$$

$$\left.\left. - \int_0^t \langle D_u L(s, X(s); b_X(s, X(s))), \sigma(s, X(s))dW_X(s)\rangle\right|^2\right]$$

$$\leq 4E\left[\int_0^1 |\sigma(s, X(s))^t(D_x\varphi_n(s, X(s)) - D_u L(s, X(s); b_X(s, X(s))))|^2 ds\right]$$

$$\leq 4CE\left[\int_0^1 \{L(s, X(s); b_X(s, X(s))) - \langle b_X(s, X(s)), D_x\varphi_n(s, X(s))\rangle + H(s, X(s); D_x\varphi_n(s, X(s)))\}ds\right]$$

$$\to 0 \quad \text{as } n \to \infty$$

from (2.44), where

$$C := 2\sup\{\langle \sigma(t,x)^t D_u^2 L(t,x;u)^{-1}\sigma(t,x)z, z\rangle | t \in [0,1], x, u, z \in \mathbb{R}^d, |z| = 1\}.$$

Indeed, for a twice differentiable, strictly convex function $f : \mathbb{R}^d \longrightarrow [0, \infty)$ for which $f(u)/|u| \to \infty$ as $|u| \to \infty$ and $(u, z) \in \mathbb{R}^d \times \mathbb{R}^d$, by Taylor's Theorem, there exists $\theta \in (0, 1)$ such that

$$\begin{aligned} &f(u) - \{\langle u, z\rangle - f^*(z)\} \\ &= f^*(z) - f^*(Df(u)) - \langle Df^*(Df(u)), z - Df(u)\rangle \\ &= \frac{1}{2}\langle D^2 f^*(Df(u) + \theta(z - Df(u)))(z - Df(u)), z - Df(u)\rangle, \end{aligned}$$

and $D^2 f^*(z) = D^2 f(Df^*(z))^{-1}$ (see [152, 2.1.3]). Here

$$f^*(z) := \sup\{\langle z, u\rangle - f(u) | u \in \mathbb{R}^d\}.$$

From (2.44), (2.111), and (2.112), $\varphi_n(1, y) - \varphi_n(0, x)$ is convergent in $L^1(\mathbb{R}^d \times \mathbb{R}^d, P^{(X(0),X(1))})$.

From (A.4) and (A.5), $P^{X(\cdot)}$ is absolutely continuous with respect to $P^{X^0(\cdot)}$ (see (2.66) for notation). Indeed, by Taylor's Theorem, there exists $\theta \in (0, 1)$ such that

$$L(t, x; u) = L(t, x; 0) + \langle D_u L(t, x; 0), u\rangle + \frac{1}{2}\langle D_u^2 L(t, x; \theta u)u, u\rangle.$$

In particular, since X^0 has a transition probability density (see, e.g. [66]),

$$p(t, y) := \frac{P(X(t) \in dy)}{dy} \text{ exists}, \quad t \in (0, 1],$$

$$p(0, x; t, y) := \frac{P(X(t) \in dy | X(0) = x)}{dy} \text{ exists } P_0(dx)\text{–a.e.}, \quad t \in (0, 1].$$

Therefore $P^{(X(0),X(1))}(dxdy)$ is absolutely continuous with respect to $P_0(dx)$ $P_1(dy)$. Indeed,

$$P^{(X(0),X(1))}(dxdy) = \frac{p(0,x;1,y)}{p(1,y)} P_0(dx)P_1(dy).$$

Hence, from [135, Proposition 2], there exist measurable functions f_i on $\mathbb{R}^d$, $i = 0, 1$ such that

$$\lim_{n\to\infty} E[|\varphi_n(1, X(1)) - \varphi_n(0, X(0)) - \{f_1(X(1)) - f_0(X(0))\}|] = 0. \tag{2.113}$$

$$Y(t) := f_0(X(0)) + \int_0^t L(s, X(s); b_X(s, X(s)))ds \tag{2.114}$$

$$+ \int_0^t \langle D_u L(s, X(s); b_X(s, X(s))), \sigma(s, X(s))dW_X(s)\rangle.$$

From Lemma 2.6 and (2.111)–(2.113), (2.46) holds. □

We can prove Corollary 2.3 in exactly the same way as in Theorem 2.5.

Proof of Corollary 2.3 The Duality Theorem holds under (A.4)'. Indeed, (A.4)' implies that (A.0)–(A.3) hold and that the unique classical solution to the HJB Eqn. (2.39) is also the minimal bounded continuous viscosity solution to the HJB Eqn. (2.39) if $\varphi(1,\cdot) \in C_b^3(\mathbb{R}^d)$, as we explained before we stated Corollary 2.3 (see also Lemma 2.5). In particular, Corollary 2.2 holds under (A.4)'. (A.4)' implies (A.5) and the proof of Theorem 2.5 can be applied even though (A.4) does not hold. In particular,

$$f_1(X(1)) - f_0(X(0)) - \int_0^1 c(s, X(s))ds \tag{2.115}$$

$$= \int_0^1 \langle a(s, X(s))^{-1}(b_X(s, X(s)) - \xi(s, X(s))), dX(s) - \xi(s, X(s))ds >$$

$$-\frac{1}{2}\int_0^1 |\sigma(s, X(s))^{-1}(b_X(s, X(s)) - \xi(s, X(s)))|^2 ds,$$

which completes the proof (see [100]). □

2.2.3 *Finiteness of SOT*

In this section, we give a sufficient condition under which $V(P_0, P_1)$ is finite, which implies the existence of a semimartingale with a given diffusion matrix and initial and terminal distributions. The idea of the proof is to show the finiteness of the supremum in the Duality Theorem for $V(P_0, P_1)$. Our result also gives an approach, in the framework of "stochastic optimal transportation", to the construction of an h-path process for Brownian motion with given initial and terminal distributions, in that we do not solve Schrödinger's Func. Eqn. Instead, we only prove the finiteness of and the Duality Theorem for $V(P_0, P_1)$. The result in this section is from [115] and completes the program of this approach, though Jamison's result [80] doesn't assume any integrality condition.

The following are assumptions:

(C.1). $p(t,x) := P_t(dx)/dx$ exists and is absolutely continuous in x for $t = 0, 1$.
(C.2). The following is finite: for $\mu(dxdy) := p(0,x)p(1,y)dxdy$,

$$\begin{aligned}
&\int_0^1 dt \int_{\mathbb{R}^d\times\mathbb{R}^d} L(t,(1-t)x+ty;3(y-x))\mu(dxdy) \qquad (2.116)\\
&+\int_0^1 \min\Bigg(\int_{\mathbb{R}^d\times\mathbb{R}^d} L\bigg(t,(1-t)x+ty;\\
&\qquad \frac{1}{1-t}a(t,(1-t)x+ty)D_x \log p(0,x)\bigg)\mu(dxdy),\\
&\qquad \int_{\mathbb{R}^d\times\mathbb{R}^d} L\bigg(t,(1-t)x+ty;\\
&\qquad \frac{1}{t}a(t,(1-t)x+ty)D_y \log p(1,y)\bigg)\mu(dxdy)\Bigg)dt.
\end{aligned}$$

Let $\mathscr{V}_{cla}(P_0, P_1)$ denote the supremum in the Duality Theorem in Corollary 2.2, i.e.,

$$\mathscr{V}_{cla}(P_0,P_1) := \sup\left\{\int_{\mathbb{R}^d}\varphi(1,x)P_1(dx) - \int_{\mathbb{R}^d}\varphi(0,x)P_0(dx)\right\}, \qquad (2.117)$$

where the supremum is taken over all classical solutions φ to the HJB Eqn. (2.39).

Theorem 2.7 (See [115]) *Suppose that (A.2) and (A.4) hold. Then for any $P_0, P_1 \in \mathscr{P}(\mathbb{R}^d)$ such that (C.1)–(C.2) hold, $\mathscr{V}_{cla}(P_0, P_1)$ is finite.*

The following gives a sufficient condition for the finiteness of $V(\cdot,\cdot)$. It can be obtained easily from Corollary 2.2 and Theorem 2.7 and the proof is omitted.

Corollary 2.4 (See [115]) *Suppose that (A.2), (A.3,ii), and (A.4) hold. Then for any $P_0, P_1 \in \mathscr{P}(\mathbb{R}^d)$ such that (C.1)–(C.2) hold, $V(P_0, P_1)$ is finite, a minimizer $\{X(t)\}_{0\le t\le 1}$ of $V(P_0, P_1)$ exists and (2.45) holds.*

Before we state a corollary to Corollary 2.4, we introduce a stronger and simpler assumption than (C.2):

(C.2)'. There exists $r > 1$ such that the following holds:

$$\int_{\mathbb{R}^d} (|x|^r + |D_x \log p(t, x)|^r) p(t, x) dx < \infty, \quad t = 0, 1. \tag{2.118}$$

Applying Corollary 2.4 to $L = (1 + |u|^2)^{\frac{r}{2}}$ for $r > 1$, the following holds.

Corollary 2.5 (See [115]) *Suppose that σ is uniformly nondegenerate and $\sigma_{ij} \in C_b^{1,2}([0, 1] \times \mathbb{R}^d)$, $i, j = 1, \cdots, d$. Then for any P_0, $P_1 \in \mathcal{P}(\mathbb{R}^d)$ such that (C.1) and (C.2)' hold, there exists $\{X(t)\}_{0 \le t \le 1} \in \mathcal{A}$ such that the following holds:*

$$\begin{aligned} dX(t) &= b_X(t, X(t))dt + \sigma(t, X(t))dW_X(t), \quad 0 \le t \le 1, \qquad (2.119)\\ PX(t)^{-1} &= P_t, \quad t = 0, 1, \\ & E\left[\int_0^1 |b_X(t, X(t))|^r dt\right] < +\infty. \end{aligned}$$

In particular, if $r = 2$, then there uniquely exists the h-path process for Brownian motion on the time interval $[0, 1]$ *with initial and terminal distributions P_0 and P_1, respectively.*

Let x^t denote a transpose of a vector x. Then

$$\{D_x^t a(t, x)\}^t = \left(\sum_{i=1}^d \frac{\partial a_{ij}(t, x)}{\partial x_i}\right)_{j=1}^d.$$

We state the key lemma for the proof of Theorem 2.7.

Lemma 2.7 *Suppose that (A.0)–(A.2) hold. Then for any P_0, $P_1 \in \mathcal{P}(\mathbb{R}^d)$ such that (C.1)–(C.2) hold, the following holds: for any solution $\varphi(t, x) \in C_b^{1,2}([0, 1] \times \mathbb{R}^d)$ of the HJB Eqn. (2.39),*

$$\begin{aligned} &\int_{\mathbb{R}^d} \varphi(1, y)p(1, y)dy - \int_{\mathbb{R}^d} \varphi(0, x)p(0, x)dx \qquad (2.120)\\ &\le \frac{1}{2}\int_0^1 dt \int_{\mathbb{R}^d \times \mathbb{R}^d} L(t, (1 - t)x + ty; \\ &\qquad 2(y - x) + \{D_z^t a(t, (1 - t)x + ty)\}^t)\mu(dxdy) \\ &+ \frac{1}{2}\int_0^1 \min\left(\int_{\mathbb{R}^d \times \mathbb{R}^d} L\Big(t, (1 - t)x + ty; \right. \end{aligned}$$

$$\frac{1}{1-t}a(t,(1-t)x+ty)D_x \log p(0,x)\Big)\mu(dxdy),$$

$$\int_{\mathbb{R}^d\times\mathbb{R}^d} L\Big(t,(1-t)x+ty;$$

$$\frac{1}{t}a(t,(1-t)x+ty)D_y \log p(1,y)\Big)\mu(dxdy)\Big)dt.$$

Proof of Corollary 2.3 On a Borel probability space $(\mathbb{R}^d \times \mathbb{R}^d, \mathbf{B}(\mathbb{R}^d \times \mathbb{R}^d), \mu)$ (see (C.2) for notation),

$$X(t) = X(t; x, y) := (1-t)x + ty, \quad (t, x, y) \in [0, 1] \times \mathbb{R}^d \times \mathbb{R}^d. \tag{2.121}$$

Denote by $p(t, x)$, $(t, x) \in (0, 1) \times \mathbb{R}^d$, the probability density function of $X(t)$, i.e.,

$$\begin{aligned} p(t,z) &:= \int_{\mathbb{R}^d} p(1,y)\frac{1}{(1-t)^d}p\Big(0,\frac{z-ty}{1-t}\Big)dy \\ &= \int_{\mathbb{R}^d} p(0,x)\frac{1}{t^d}p\Big(1,\frac{z-(1-t)x}{t}\Big)dx, \quad (t,z) \in (0,1)\times\mathbb{R}^d. \end{aligned} \tag{2.122}$$

Then, for any solution $\varphi(t, x) \in C_b^{1,2}([0, 1] \times \mathbb{R}^d)$ of the HJB Eqn. (2.39),

$$\begin{aligned} &\int_{\mathbb{R}^d} \varphi(1,y)p(1,y)dy - \int_{\mathbb{R}^d} \varphi(0,x)p(0,x)dx \\ &= \int_0^1 E\Big[\langle D_x\varphi(t,X(t)), X(1)-X(0)\rangle + \frac{\partial \varphi(t,X(t))}{\partial t}\Big]dt \\ &= \int_0^1 E\Big[\langle D_x\varphi(t,X(t)), X(1)-X(0)\rangle - \frac{1}{2}\langle D_x^2\varphi(t,X(t)), a(t,X(t))\rangle \\ &\qquad - H(t,X(t); D_x\varphi(t,X(t)))\Big]dt \\ &= \int_0^1 E\Big[\Big\langle D_x\varphi(t,X(t)), X(1)-X(0) + \frac{\{D_x^t(a(t,X(t))p(t,X(t)))\}^t}{2p(t,X(t))}\Big\rangle \\ &\qquad - H(t,X(t); D_x\varphi(t,X(t)))\Big]dt \\ &\le \int_0^1 E\Big[L\Big(t,X(t); X(1)-X(0) + \frac{\{D_x^t(a(t,X(t))p(t,X(t)))\}^t}{2p(t,X(t))}\Big)\Big]dt \end{aligned} \tag{2.123}$$

$$\leq \frac{1}{2}\int_0^1 E[L(t, X(t); 2(X(1) - X(0)) + \{D_x^t a(t, X(t))\}^t)$$
$$+ L(t, X(t); a(t, X(t))D_x \log p(t, X(t)))]dt.$$

We explain why the last equality and the following inequalities in (2.123) hold. Since $D_x \log p(t, x) \in L^1(p(t, x)dx)$ for $t = 0, 1$ from (A.0), (A.2), and (C.1)–(C.2), $D_x \log p(t, x) \in L^1(p(t, x)dtdx)$ and the last equality in (2.123) is true. Indeed, from (2.122), for $t \in (0, 1)$,

$$D_z p(t, z) \tag{2.124}$$
$$= \int_{\mathbb{R}^d} \left\{ D_z \log p\left(0, \frac{z - ty}{1 - t}\right)\right\} p(1, y) \frac{1}{(1-t)^d} p\left(0, \frac{z - ty}{1 - t}\right) dy,$$
$$= \int_{\mathbb{R}^d} \left\{ D_z \log p\left(1, \frac{z - (1-t)x}{t}\right)\right\} p(0, x) \frac{1}{t^d} p\left(1, \frac{z - (1-t)x}{t}\right) dx,$$

dz–a.e. Equation (2.124) implies the following:

$$\int_0^1 dt \int_{\mathbb{R}^d} |D_z p(t, z)| dz \tag{2.125}$$
$$\leq \int_0^{\frac{1}{2}} \frac{1}{1-t} dt \int_{\mathbb{R}^d \times \mathbb{R}^d} p(1, y) |D_x p(0, x)| dx dy$$
$$+ \int_{\frac{1}{2}}^1 \frac{1}{t} dt \int_{\mathbb{R}^d \times \mathbb{R}^d} p(0, x) |D_y p(1, y)| dx dy$$
$$= \log 2 \int_{\mathbb{R}^d} (|D_x p(0, x)| + |D_x p(1, x)|) dx < \infty.$$

The first inequality in (2.123) is true from (2.110). The second inequality in (2.123) is true. Indeed, $u \mapsto L(t, x; u)$ is convex from (A.1,ii) and

$$\{D_x^t (a(t, x) p(t, x))\}^t = \left(\sum_{i=1}^d \frac{\partial (a_{ij}(t, x) p(t, x))}{\partial x_i}\right)_{j=1}^d$$
$$= \left(\sum_{i=1}^d \frac{\partial a_{ij}(t, x)}{\partial x_i}\right)_{j=1}^d p(t, x) + a(t, x) D_x p(t, x)$$

since $a(t, x)$ is symmetric.

The following completes the proof: for $t \in (0, 1)$,

$$
\begin{aligned}
&E[L(t, X(t); a(t, X(t))D_x \log p(t, X(t)))] \qquad (2.126)\\
&\le \begin{cases} \displaystyle\int_{\mathbb{R}^d \times \mathbb{R}^d} L\left(t, z; a(t,z)D_z \log p\left(0, \frac{z-ty}{1-t}\right)\right) \\ \qquad \times\, p(1,y)\dfrac{1}{(1-t)^d} p\left(0, \dfrac{z-ty}{1-t}\right) dydz, \\ \displaystyle\int_{\mathbb{R}^d \times \mathbb{R}^d} L\left(t, z; a(t,z)D_z \log p\left(1, \frac{z-(1-t)x}{t}\right)\right) \\ \qquad \times\, p(0,x)\dfrac{1}{t^d} p\left(1, \dfrac{z-(1-t)x}{t}\right) dxdz. \end{cases}
\end{aligned}
$$

Equation (2.126) is true from (A.1,ii). Indeed, from (2.122) and (2.124),

$$
D_z \log p(t,z) = \begin{cases} \dfrac{\displaystyle\int_{\mathbb{R}^d} \left\{ D_z \log p\left(0, \frac{z-ty}{1-t}\right)\right\} p(1,y)\frac{1}{(1-t)^d} p\left(0, \frac{z-ty}{1-t}\right) dy}{\displaystyle\int_{\mathbb{R}^d} p(1,y)\frac{1}{(1-t)^d} p\left(0, \frac{z-ty}{1-t}\right) dy}, \\[2ex] \dfrac{\displaystyle\int_{\mathbb{R}^d} \left\{ D_z \log p\left(1, \frac{z-(1-t)x}{t}\right)\right\} p(0,x)\frac{1}{t^d} p\left(1, \frac{z-(1-t)x}{t}\right) dx}{\displaystyle\int_{\mathbb{R}^d} p(0,x)\frac{1}{t^d} p\left(1, \frac{z-(1-t)x}{t}\right) dx}. \end{cases}
$$

By Jensen's inequality,

$$
\begin{aligned}
&L(t, z; a(t,z)D_z \log p(t,z))\\
&\le \begin{cases} \dfrac{1}{p(t,z)} \displaystyle\int_{\mathbb{R}^d} L\left(t, z; a(t,z)D_z \log p\left(0, \frac{z-ty}{1-t}\right)\right) \\ \qquad \times\, p(1,y)\dfrac{1}{(1-t)^d} p\left(0, \dfrac{z-ty}{1-t}\right) dy, \\ \dfrac{1}{p(t,z)} \displaystyle\int_{\mathbb{R}^d} L\left(t, z; a(t,z)D_z \log p\left(1, \frac{z-(1-t)x}{t}\right)\right) \\ \qquad \times\, p(0,x)\dfrac{1}{t^d} p\left(1, \dfrac{z-(1-t)x}{t}\right) dx. \end{cases}
\end{aligned}
$$

□

We prove Theorem 2.7.

Proof of Theorem 2.7 On the right-hand side of (2.120), from (A.4, ii),

$$
\begin{aligned}
&L(t, (1-t)x + ty; 2(y-x) + \{D_z^t a(t, (1-t)x+ty)\}^t) \qquad (2.127)\\
&\le \frac{2}{3} L(t, (1-t)x+ty; 3(y-x))\\
&\qquad + \frac{1}{3} L(t, (1-t)x+ty; 3\{D_z^t a(t, (1-t)x+ty)\}^t).
\end{aligned}
$$

$L(t,(1-t)x+ty;\{D_z^t a(t,(1-t)x+ty)\}^t)$ is bounded from (A.4,iii), since $D_z^t a(t,z)$ is bounded from (A.4,i). In particular, $\mathscr{V}_{cla}(P_0,P_1)$ is finite from Lemma 2.7 since (A.4,ii) implies (A.1) and since the set over which the supremum is taken in $\mathscr{V}_{cla}(P_0,P_1)$ is not empty (see the explanation below (2.43)). □

We prove Corollary 2.5.

Proof of Corollary 2.5 Set $L = (1+|u|^2)^{\frac{r}{2}}$. Then (A.2)–(A.4) are satisfied and (2.116) can be rewritten as follows:

$$\int_0^1 dt \int_{\mathbb{R}^d\times\mathbb{R}^d} (1+|3(y-x)|^2)^{\frac{r}{2}}\mu(dxdy) \tag{2.128}$$
$$+\int_0^1 \min\left(\int_{\mathbb{R}^d\times\mathbb{R}^d}\left(1+\left|\frac{1}{1-t}a(t,(1-t)x+ty)D_x\log p(0,x)\right|^2\right)^{\frac{r}{2}}\mu(dxdy),\right.$$
$$\left.\int_{\mathbb{R}^d\times\mathbb{R}^d}\left(1+\left|\frac{1}{t}a(t,(1-t)x+ty)D_y\log p(1,y)\right|^2\right)^{\frac{r}{2}}\mu(dxdy)\right)dt.$$

The following (2.129)–(2.130) imply that (C.2) holds from (C.2').

$$(1+|3(y-x)|^2)^{\frac{r}{2}} \le 2^{\frac{r}{2}}(1+3^r|y-x|^r) \le 2^{\frac{r}{2}}(1+6^r(|y|^r+|x|^r)), \tag{2.129}$$

since for any $x, y \ge 0$,

$$(x+y)^{\frac{r}{2}} \le (2\max(x,y))^{\frac{r}{2}} \le 2^{\frac{r}{2}}(x^{\frac{r}{2}}+y^{\frac{r}{2}}).$$

In the same way, we have

$$\int_0^1 \min\left(\int_{\mathbb{R}^d\times\mathbb{R}^d}\left(1+\left|\frac{1}{1-t}a(t,(1-t)x+ty)D_x\log p(0,x)\right|^2\right)^{\frac{r}{2}}\mu(dxdy),\right.$$
$$\left.\int_{\mathbb{R}^d\times\mathbb{R}^d}\left(1+\left|\frac{1}{t}a(t,(1-t)x+ty)D_y\log p(1,y)\right|^2\right)^{\frac{r}{2}}\mu(dxdy)\right)dt \tag{2.130}$$
$$\le \int_0^{\frac{1}{2}} dt \int_{\mathbb{R}^d\times\mathbb{R}^d} (1+|2a(t,(1-t)x+ty)D_x\log p(0,x)|^2)^{\frac{r}{2}}\mu(dxdy)$$
$$+\int_{\frac{1}{2}}^1 dt \int_{\mathbb{R}^d\times\mathbb{R}^d} (1+|2a(t,(1-t)x+ty)D_y\log p(1,y)|^2)^{\frac{r}{2}}\mu(dxdy)$$
$$\le 2^{\frac{r}{2}}\left(1+2^{r-1}\sup_{(t,z)\in[0,1]\times\mathbb{R}^d}|a(t,z)|^r\int_{\mathbb{R}^d\times\mathbb{R}^d}|D_x\log p(0,x)|^r\mu(dxdy)\right.$$
$$\left.+2^{r-1}\sup_{(t,z)\in[0,1]\times\mathbb{R}^d}|a(t,z)|^r\int_{\mathbb{R}^d\times\mathbb{R}^d}|D_y\log p(1,y)|^r\mu(dxdy)\right),$$

which completes the proof. □

2.3 Zero-Noise Limit of SOT

In this section, we show that one can solve Monge's problem with a quadratic cost by the zero-noise limit of the SOT, the minimizer of which is an h-path process. We also show that the zero-noise limit of the Duality Theorem for the SOT gives the Duality Theorem for the OT. The purpose of such research is that we would like to study the OT in the framework of the SOT as a problem of the principle of least action for particle systems.

2.3.1 *Monge's Problem by the Zero-Noise Limit of SOT*

By the zero-noise limit of h-path processes, we prove the existence and the uniqueness of the minimizer of Monge's problem with a quadratic cost, independently of Kantorovich's approach (see [108]). In this sense, the SOT can be also considered the OT with a random external force.

For $\varepsilon > 0$ and $P_1 \in \mathscr{P}(\mathbb{R}^d)$,

$$P_{1,\varepsilon}(dy) := g(\varepsilon, \cdot) * P_1(dy) = \left(\int_{\mathbb{R}^d} g(\varepsilon, y - z) P_1(dz) \right) dy, \tag{2.131}$$

where g denotes a Gaussian kernel (see (1.47)).

Let $\{X_\varepsilon(t)\}_{0 \le t \le 1}$ be an h-path process in Corollary 2.3 in the case where $\sigma(t, x) = \sqrt{\varepsilon} \times \mathrm{Id}$, $\xi \equiv 0$, $c \equiv 0$, and P_1 is replaced by $P_{1,\varepsilon}$ and let h_ε denote the h-function (see also Theorems 1.2 and 3.1, and [80]). That is, there exist nonnegative σ-finite Borel measures $\nu_{1,\varepsilon}(dx)$ and $\nu_{2,\varepsilon}(dx) = \nu_{2,\varepsilon}(x)dx$ such that the following holds in a weak sense:

$$\begin{aligned} dX_\varepsilon(t) &= \varepsilon D_x \log h_\varepsilon(t, X_\varepsilon(t))dt + \sqrt{\varepsilon} dW_{X_\varepsilon}(t), \quad 0 < t < 1, \\ P^{X_\varepsilon(0)} &= P_0, \quad P^{X_\varepsilon(1)} = P_{1,\varepsilon}, \\ \mu_\varepsilon(dxdy) &:= P^{(X_\varepsilon(0), X_\varepsilon(1))}(dxdy) = \nu_{1,\varepsilon}(dx) g(\varepsilon, y - x) \nu_{2,\varepsilon}(x) dx, \\ h_\varepsilon(t, x) &:= \begin{cases} \displaystyle\int_{\mathbb{R}^d} g(\varepsilon(1-t), x - y) \nu_{2,\varepsilon}(y) dy, & (t, x) \in [0, 1) \times \mathbb{R}^d, \\ \nu_{2,\varepsilon}(x), \quad (t, x) \in \{1\} \times \mathbb{R}^d. \end{cases} \end{aligned} \tag{2.132}$$

Here one can assume that $X_0 := X_\varepsilon(0)$ is independent of ε. We write $V_{S,\varepsilon} := V$ in the case where $L = (2\varepsilon)^{-1}|u|^2$ under the setting stated above. Then from Corollary 2.3 (see also [156]), $\{X_\varepsilon(t)\}_{0 \le t \le 1}$ is the unique minimizer of $V_{S,\varepsilon}(P_0, P_{1,\varepsilon})$, provided it is finite.

It seems likely that the h-path process $\{X_\varepsilon(t)\}_{0 \le t \le 1}$ converges, as $\varepsilon \to 0$, to the minimizer of Monge's problem $T_M(P_0, P_1)$ in (2.7) with $c(x, y) = |y - x|^2$ (see

(1.31)). But it is not trivial that the limit is a function of t and X_0 since a continuous strong Markov process which is of bounded variation in time is not necessarily a function of the initial point and time (see [137] and also [107]). Independently of known results on the Monge–Kantorovich problem, we show that $2\varepsilon V_{S,\varepsilon}(P_0, P_{1,\varepsilon})$ converges to $T_M(P_0, P_1)$ and that $X_\varepsilon(1)$ is convergent in L^2 and the limit is the minimizer of $T_M(P_0, P_1)$ with $c(x, y) = |y - x|^2$, as $\varepsilon \to 0$. This a probabilistic proof of the existence of the minimizer of $T_M(P_0, P_1)$ with $c(x, y) = |y - x|^2$.

We recall the definition of cyclic monotonicity of a set.

Definition 2.6 A nonempty set $A \in \mathbb{R}^d \times \mathbb{R}^d$ is called cyclically monotone if for any $n \geq 1$ and any $(x_i, y_i) \in A, i = 1, \cdots, n$,

$$\sum_{i=1}^{n} \langle y_i, x_{i+1} - x_i \rangle \leq 0 \tag{2.133}$$

(see, e.g. [152, p. 80]), where $x_{n+1} := x_1$.

$$\mathscr{P}_2(\mathbb{R}^d) := \left\{ P \in \mathscr{P}(\mathbb{R}^d) \middle| \int_{\mathbb{R}^d} |x|^2 P(dx) < \infty \right\}, \tag{2.134}$$

$$\mathscr{P}_{ac}(\mathbb{R}^d) := \{\rho(x)dx \in \mathscr{P}(\mathbb{R}^d)\}, \tag{2.135}$$

$$\mathscr{P}_{2,ac}(\mathbb{R}^d) := \mathscr{P}_2(\mathbb{R}^d) \cap \mathscr{P}_{ac}(\mathbb{R}^d). \tag{2.136}$$

Then the following holds.

Theorem 2.8 (See [108]) *For any $P_0, P_1 \in \mathscr{P}_2(\mathbb{R}^d)$, $\{\mu_\varepsilon\}_{\varepsilon \in (0,1]}$ is tight and any weak limit point of $\{\mu_\varepsilon\}_{\varepsilon \in (0,1]}$, as $\varepsilon \to 0$, is supported on a cyclically monotone set.*

Since a cyclically monotone set in $\mathbb{R}^d \times \mathbb{R}^d$ is contained in the subdifferential of a proper lower semicontinuous convex function on $\mathbb{R}^d$ and since a proper convex function is differentiable dx–a.e. in the interior of its domain (see [152, pp. 52, 82]), we obtain the following.

Corollary 2.6 (See [108]) *For any $P_0 \in \mathscr{P}_{2,ac}(\mathbb{R}^d)$ and $P_1 \in \mathscr{P}_2(\mathbb{R}^d)$ and for any weak limit point μ of $\{\mu_\varepsilon\}_{\varepsilon \in (0,1]}$ as $\varepsilon \to 0$, there exists a proper lower semicontinuous convex function $\varphi : \mathbb{R}^d \longrightarrow (-\infty, \infty]$ such that*

$$\mu(dxdy) = P_0(dx)\delta_{D\varphi(x)}(dy). \tag{2.137}$$

Remark 2.10 For any $P_0, P_1 \in \mathscr{P}_{2,ac}(\mathbb{R}^d)$, Corollary 2.6 gives a probabilistic proof of the existence of a solution to the following Monge–Ampère equation:

$$p(0, x) = p(1, D\varphi(x)) \det(D^2\varphi(x)) \tag{2.138}$$

in the sense that $P_0(D\varphi)^{-1} = P_1$ (see (C.1) for notation). For the regularity results on the solution to (2.138), see [152, pp. 140–141], [9, Theorem 1.1] and the references therein.

$$b_\varepsilon(t,x) := \varepsilon D_x \log h_\varepsilon(t,x) \tag{2.139}$$

(see (2.132) for notation). The following which can be proved from Theorem 2.8 and Corollary 2.6, independently of known results on the OT (see [4, 21, 22, 88, 130, 144, 152]) is our main result in this section.

Theorem 2.9 (See [108]) *Suppose that $c(x,y) = |y-x|^2$. Then for any $P_0 \in \mathscr{P}_{2,ac}(\mathbb{R}^d)$ and $P_1 \in \mathscr{P}_2(\mathbb{R}^d)$,*

$$\lim_{\varepsilon\to 0} 2\varepsilon V_{S,\varepsilon}(P_0, P_{1,\varepsilon}) = T_M(P_0, P_1) < \infty. \tag{2.140}$$

In particular, $D\varphi$ in Corollary 2.6 is the unique minimizer of $T_M(P_0,P_1)$, and the following holds:

$$\lim_{\varepsilon\to 0} E\left[\int_0^1 |b_\varepsilon(t, X_\varepsilon(t)) - (D\varphi(X_0) - X_0)|^2 dt\right] = 0, \tag{2.141}$$

$$\lim_{\varepsilon\to 0} E\left[\sup_{0\le t\le 1} |X_\varepsilon(t) - \{X_0 + t(D\varphi(X_0) - X_0)\}|^2\right] = 0. \tag{2.142}$$

As far as we know, the following is not known.

Corollary 2.7 *Suppose that $c(x,y) = |y-x|^2$. Then for any $P_0 \in \mathscr{P}_{2,ac}(\mathbb{R}^d)$, $P_1 \in \mathscr{P}_2(\mathbb{R}^d)$, and $D\varphi$ in Corollary 2.6, the following holds: for $t \in [0,1)$*

$$\lim_{\varepsilon\to 0} E\left[\sup_{0\le s\le t} |b_\varepsilon(s, X_\varepsilon(s)) - (D\varphi(X_0) - X_0)|^2\right] = 0. \tag{2.143}$$

Remark 2.11 If $h_\varepsilon(t,x) \in C^{1,3}([0,1)\times\mathbb{R}^d)$, then from Theorem 3.2 in Sect. 3.1.1,

$$\partial_t b_\varepsilon(t,x) + \frac{\varepsilon}{2}\Delta b_\varepsilon(t,x) + D_x b_\varepsilon(t,x) b_\varepsilon(t,x) = 0, \quad (t,x) \in [0,1)\times\mathbb{R}^d, \tag{2.144}$$

where $D_x f(t,x) := (\partial f_i(t,x)/\partial x_j)_{i,j=1}^d$ for $f(t,x) := (f_i(t,x))_{i=1}^d$. Equation (2.144) implies that $\{b_\varepsilon(t, X_\varepsilon(t))\}_{0\le t<1}$ is an $(\mathscr{F}_t^{X_\varepsilon})$-local martingale, which we prove by a different approach in the proof of Corollary 2.7. It also implies that a drift vector of an h-path process is not necessarily a martingale in the case where $a(t,x)$ is not a constant matrix.

The following is known on the OT with $c(x, y) = |y - x|^2$ though it is not used in the proof of Theorem 2.9.

(i) Suppose that $P_0, P_1 \in \mathscr{P}_2(\mathbb{R}^d)$. Then a probability measure supported on a cyclically monotone set in $\mathbb{R}^d \times \mathbb{R}^d$ is a minimizer of $T(P_0, P_1)$ (see [88, 144] and also [152, pp. 66, 82], [4, Theorem 3.2]]).
(ii) Suppose that $P_0 \in \mathscr{P}_{2,ac}(\mathbb{R}^d)$ and $P_1 \in \mathscr{P}_2(\mathbb{R}^d)$. Then there exists a convex function φ such that $P_0(dx)\delta_{D\varphi(x)}(dy)$ is the unique minimizer of $T(P_0, P_1)$ (see [21, 22]).

Using these facts, we have the following from Theorem 2.8 and Corollary 2.6.

Corollary 2.8 (See [108]) *Suppose that* $c(x, y) = |y - x|^2$. *Then (i) for any* $P_0, P_1 \in \mathscr{P}_2(\mathbb{R}^d)$, *any weak limit point of* $\{\mu_\varepsilon\}_{\varepsilon \in (0,1]}$ *as* $\varepsilon \to 0$ *is a minimizer of* $T(P_0, P_1)$. *(ii) For any* $P_0 \in \mathscr{P}_{2,ac}(\mathbb{R}^d)$ *and* $P_1 \in \mathscr{P}_2(\mathbb{R}^d)$, μ_ε *weakly converges to the unique minimizer of* $T(P_0, P_1)$, *as* $\varepsilon \to 0$.

We considered $P_{1,\varepsilon}$, instead of P_1, to avoid technical assumptions in Theorem 2.9. As far as the zero-noise limit of h-path processes is concerned, there is no reason to perturb the terminal distribution P_1. From a probabilistic interest, we discuss the zero-noise limit of h-path processes for Brownian motion with the terminal distribution P_1.

Replace $P_{1,\varepsilon}$ by P_1 above. Instead, if we assume that $P_1 \in \mathscr{P}_{2,ac}(\mathbb{R}^d)$, then one can define $\overline{\mu}_\varepsilon$, $\overline{h}_\varepsilon(t, x)$, $\overline{X}_\varepsilon(t)$ and $\overline{b}_\varepsilon(s, x)$ in the same way as in Theorem 2.9 (see (2.132)).

$$\mathscr{S}(P) := \begin{cases} \int_{\mathbb{R}^d} \{\log p(x)\} p(x) dx, & P(dx) = p(x)dx, \\ \infty, & \mathit{otherwise}. \end{cases} \tag{2.145}$$

In addition to the condition $P_0, P_1 \in \mathscr{P}_2(\mathbb{R}^d)$, assume that $\mathscr{S}(P_1)$ is finite. Then $V_{S,\varepsilon}(P_0, P_1)$ is finite for $\varepsilon > 0$ (see Proposition 3.3) and a result similar to Theorem 2.9 and Corollary 2.8 holds for $\overline{X}_\varepsilon$. More precisely, the following, the proof of which we omit, holds. We refer readers to [108].

Proposition 2.4 (See [108]) *Suppose that* $c(x, y) = |y - x|^2$. *Then (i) for any* $P_0, P_1 \in \mathscr{P}_2(\mathbb{R}^d)$, $\{\overline{\mu}_\varepsilon\}_{\varepsilon \in (0,1]}$ *is tight and any weak limit point of* $\{\overline{\mu}_\varepsilon\}_{\varepsilon \in (0,1]}$ *as* $\varepsilon \to 0$ *is a minimizer of* $T(P_0, P_1)$. *(ii) For any* $P_0 \in \mathscr{P}_{2,ac}(\mathbb{R}^d)$ *and* $P_1 \in \mathscr{P}_2(\mathbb{R}^d)$, $\overline{\mu}_\varepsilon$ *weakly converges to the unique minimizer of* $T(P_0, P_1)$, *as* $\varepsilon \to 0$. *(iii) For any* $P_0 \in \mathscr{P}_{2,ac}(\mathbb{R}^d)$ *and* $P_1 \in \mathscr{P}_2(\mathbb{R}^d)$ *for which* $\mathscr{S}(P_1)$ *is finite,* $V_{S,\varepsilon}(P_0, P_1)$ *is finite for* $\varepsilon > 0$,

$$\lim_{\varepsilon \to 0} 2\varepsilon V_{S,\varepsilon}(P_0, P_1) = T_M(P_0, P_1) \tag{2.146}$$

and for $D\varphi$ in Corollary 2.6,

$$\lim_{\varepsilon\to 0} E\left[\sup_{0\le t\le 1}|\overline{X}_\varepsilon(t)-\{X_0+t(D\varphi(X_0)-X_0)\}|^2\right]=0. \tag{2.147}$$

Remark 2.12

(i) If $P_1 \in \mathscr{P}_2(\mathbb{R}^d)$, then for $\varepsilon > 0$,

$$-\infty < \int_{\mathbb{R}^d}\{\log g(1,x)\}P_{1,\varepsilon}(dx) \le S(P_{1,\varepsilon}) \le -\log\sqrt{2\pi\varepsilon}^{\,d} < \infty \tag{2.148}$$

since the relative entropy $H(P_{1,\varepsilon}|g(1,x)dx)$ is nonnegative.

(ii) Suppose that $c(x,y) = |y-x|^2$. The following result is known under the technical assumptions on $P_0, P_1 \in \mathscr{P}_{2,ac}(\mathbb{R}^d)$ (see [1, 51, 53, 128]):

$$\lim_{\varepsilon\to 0}\left(V_{S,\varepsilon}(P_0,P_1)-\frac{1}{2\varepsilon}T(P_0,P_1)\right)=\frac{1}{2}\left(\mathscr{S}(P_1)-\mathscr{S}(P_0)\right). \tag{2.149}$$

In the rest of this section, we prove Theorems 2.8 and 2.9. We first state and prove technical lemmas.

For $x, y \in \mathbb{R}^d$, $m \ge 1$, and $\varepsilon > 0$,

$$\begin{aligned} &H_{m,\varepsilon}(x,y) &(2.150)\\ &:= \varepsilon\log\left\{\int_{U_m(o)\times U_m(o)}\exp\left(\frac{\langle x,z_2\rangle+\langle y,z_1\rangle}{\varepsilon}-\frac{\langle z_1,z_2\rangle}{\varepsilon}\right)\mu_\varepsilon(dz_1dz_2)\right\}, \end{aligned}$$

$$H_{i,m,\varepsilon}(x) := \varepsilon\log\left(\int_{U_m(o)} g(\varepsilon, x-z_j)\nu_{j,\varepsilon}(dz_j)\right)+\frac{|x|^2}{2}\ (i,j=1,2,\ i\ne j), \tag{2.151}$$

$$\mu_{1,m,\varepsilon}(dz_1) := \mu_\varepsilon(dz_1\times U_m(o)),\quad \mu_{2,m,\varepsilon}(dz_2) := \mu_\varepsilon(U_m(o)\times dz_2), \tag{2.152}$$

where $U_m(o) := \{x \in \mathbb{R}^d : |x| < m\}$. Then the following holds.

Lemma 2.8 *For any $m \ge 1$ and $\varepsilon > 0$ for which $\mu_\varepsilon(U_m(o)\times U_m(o)) > 0$, the following holds. (i) For any x and $y \in \mathbb{R}^d$,*

$$H_{m,\varepsilon}(x,y) = H_{1,m,\varepsilon}(x)+H_{2,m,\varepsilon}(y)+\varepsilon\log\sqrt{2\pi\varepsilon}^{\,d}, \tag{2.153}$$

$$\mu_\varepsilon(dz_1dz_2)=\exp\left(\frac{1}{\varepsilon}(\langle z_1,z_2\rangle - H_{m,\varepsilon}(z_1,z_2))\right)\mu_{1,m,\varepsilon}(dz_1)\mu_{2,m,\varepsilon}(dz_2), \tag{2.154}$$

$$H_{m,\varepsilon}(x, y) = \varepsilon \log \left\{ \int_{U_m(o)\times U_m(o)} \exp\left(\frac{\langle x, z_2\rangle + \langle y, z_1\rangle}{\varepsilon} - \frac{H_{m,\varepsilon}(z_1, z_2)}{\varepsilon}\right)\mu_{1,m,\varepsilon}(dz_1)\mu_{2,m,\varepsilon}(dz_2)\right\}. \quad (2.155)$$

(ii) $H_{m,\varepsilon}(\cdot,\cdot)$ *is convex, and for any x and* $y \in \mathbb{R}^d$,

$$|H_{m,\varepsilon}(x, y)| \le (|x| + |y|)m + m^2 - \varepsilon \log \mu_\varepsilon(U_m(o) \times U_m(o)). \quad (2.156)$$

Proof of Corollary 2.5 We first prove (i). Equation (2.153) can be obtained from (2.132) and (2.150)–(2.151) easily. Equation (2.154) holds from (2.132), (2.153), and from the following: for $i, j = 1, 2$ for which $i \neq j$,

$$\frac{\mu_{i,m,\varepsilon}(dz_i)}{\nu_{i,\varepsilon}(dz_i)} = \int_{U_m(o)} g(\varepsilon, z_i - z_j)\nu_{j,\varepsilon}(dz_j) = \exp\left(\frac{1}{\varepsilon}\left(H_{i,m,\varepsilon}(z_i) - \frac{|z_i|^2}{2}\right)\right). \quad (2.157)$$

Equation (2.155) can be obtained from (2.150) and (2.154) easily.

Next, we prove (ii). $H_{m,\varepsilon}(\cdot,\cdot)$ is convex since for any $\lambda \in (0, 1)$ and any (x, y), $(\tilde{x}, \tilde{y}) \in \mathbb{R}^d \times \mathbb{R}^d$,

$$\begin{aligned}
&H_{m,\varepsilon}(\lambda x + (1-\lambda)\tilde{x}, \lambda y + (1-\lambda)\tilde{y}) \\
&= \varepsilon \log \left\{ \int_{U_m(o)\times U_m(o)} \exp\left(\frac{\lambda(\langle x, z_2\rangle + \langle y, z_1\rangle - \langle z_1, z_2\rangle)}{\varepsilon}\right) \right. \\
&\qquad \left. \times \exp\left(\frac{(1-\lambda)(\langle \tilde{x}, z_2\rangle + \langle \tilde{y}, z_1\rangle - \langle z_1, z_2\rangle)}{\varepsilon}\right)\mu_\varepsilon(dz_1 dz_2)\right\} \\
&\le \lambda H_{m,\varepsilon}(x, y) + (1-\lambda)H_{m,\varepsilon}(\tilde{x}, \tilde{y})
\end{aligned}$$

by Hölder's inequality. Equation (2.156) can be obtained from (2.150) easily. □

Remark 2.13 For $x \in \mathbb{R}^d$, $m \ge 1$, $\varepsilon > 0$, and $i, j = 1, 2$, $i \neq j$,

$$H_{i,m,\varepsilon}(x) = \varepsilon \log \left(\int_{U_m(o)} \frac{1}{\sqrt{2\pi\varepsilon}^d} \exp\left(\frac{1}{\varepsilon}(\langle x, z_j\rangle - H_{j,m,\varepsilon}(z_j))\right)\mu_{j,m,\varepsilon}(dz_j)\right)$$

from (2.151) and (2.157).

Lemma 2.9 *Suppose that* $P_0, P_1 \in \mathscr{P}_2(\mathbb{R}^d)$. *Then for any sequence* $\{\varepsilon_n\}_{n\ge 1}$ *for which* $\varepsilon_n \to 0$ *as* $n \to \infty$, *there exist a subsequence* $\{\varepsilon_{n(k)}\}_{k\ge 1}$ *and* $m_0 \ge 1$ *such that* $H_{m,\varepsilon_{n(k)}}$ *is convergent in* $C(\mathbb{R}^d \times \mathbb{R}^d)$ *as* $k \to \infty$ *for all* $m \ge m_0$. *In particular,*

$$m \mapsto H_m := \lim_{k\to\infty} H_{m,\varepsilon_{n(k)}} (: \mathbb{R}^d \times \mathbb{R}^d \longrightarrow (-\infty, \infty)) \quad (2.158)$$

is nondecreasing on $\{m_0, m_0+1, \cdots\}$,

$$(x, y) \mapsto H(x, y) := \lim_{m\to\infty} H_m(x, y) \in (-\infty, \infty] \tag{2.159}$$

is convex on $\mathbb{R}^d \times \mathbb{R}^d$,

$$\langle x, y\rangle - H(x, y) \le 0, \quad (x, y) \in supp(P_0) \times supp(P_1), \tag{2.160}$$

and the following set is cyclically monotone:

$$S := \{(x, y) \in supp(P_0) \times supp(P_1) | \langle x, y\rangle = H(x, y)\}. \tag{2.161}$$

Proof of Corollary 2.5 There exist $m_0 \ge 1$ such that for any $m \ge m_0$, $\{H_{m,\varepsilon_n}\}_{n\ge1}$ is bounded in $U_{\ell+1}(o) \times U_{\ell+1}(o)$ for any $\ell \ge 1$ from (2.156) and from the following:

$$\begin{aligned}
&1 - \mu_\varepsilon(U_m(o) \times U_m(o)) \qquad\qquad (2.162)\\
&\le \frac{\int_{\mathbb{R}^d\times\mathbb{R}^d}(|x|^2 + |y|^2)\mu_\varepsilon(dxdy)}{m^2} = \frac{\int_{\mathbb{R}^d}|x|^2 P_0(dx) + \int_{\mathbb{R}^d}|y|^2 P_{1,\varepsilon}(dy)}{m^2}\\
&= \frac{\int_{\mathbb{R}^d}|x|^2 P_0(dx) + \int_{\mathbb{R}^d\times\mathbb{R}^d}|x + y|^2 g(\varepsilon, x)dx P_1(dy)}{m^2}\\
&\le \frac{\int_{\mathbb{R}^d}|x|^2 P_0(dx) + 2(\varepsilon d + \int_{\mathbb{R}^d}|y|^2 P_1(dy))}{m^2} \to 0, \quad m \to \infty,
\end{aligned}$$

uniformly for $\varepsilon = \varepsilon_n$, $n \ge 1$. Hence for any $m \ge m_0$ and any $\ell \ge 1$, $\{H_{m,\varepsilon_n}\}_{n\ge1}$ contains a uniformly convergent subsequence on $U_\ell(o) \times U_\ell(o)$ since $H_{m,\varepsilon_n}(\cdot,\cdot)$ is convex from Lemma 2.8, (ii) (see [8, p. 21, Theorem 3.2]). By the diagonal method, $\{H_{m,\varepsilon_n}\}_{n\ge1}$ contains a convergent subsequence $\{H_{m,\varepsilon_{m,n}}\}_{n\ge1}$ in $C(\mathbb{R}^d \times \mathbb{R}^d)$. In particular, we can take $\{\varepsilon_{m,n}\}_{n\ge1}$ so that $m \mapsto \{\varepsilon_{m,n}\}_{n\ge1}$ is nonincreasing on $\{m_0, m_0+1, \cdots\}$.

$$\varepsilon_{n(k)} := \varepsilon_{k+m_0-1,k+m_0-1} \quad (k \ge 1).$$

Then $H_{m,\varepsilon_{n(k)}}$ is convergent in $C(\mathbb{R}^d \times \mathbb{R}^d)$ as $k \to \infty$ for all $m \ge m_0$.

$m \mapsto H_m$ is nondecreasing on $\{m_0, m_0+1, \cdots\}$ since

$$H_{m+1,\varepsilon_{n(k)}} \ge H_{m,\varepsilon_{n(k)}}, \quad k \ge 1$$

for all $m \ge m_0$ from (2.150). Hence for any $(x, y) \in \mathbb{R}^d \times \mathbb{R}^d$, $H_m(x, y)$ is convergent or diverges to ∞ as $m \to \infty$.

As the limit of convex functions, $H(\cdot,\cdot)$ in (2.159) is convex in $\mathbb{R}^d \times \mathbb{R}^d$.

For any $(x, y) \in \operatorname{supp}(P_0) \times \operatorname{supp}(P_1)$, $r > 0$, $m \geq r + |x| + |y| + m_0$, and $k \geq 1$, from (2.155),

$$\begin{aligned} &H_{m,\varepsilon_{n(k)}}(x, y) \qquad (2.163)\\ &\geq \inf_{(z_0,z_1)\in U_r(x)\times U_r(y)} \{\langle x, z_2\rangle + \langle y, z_1\rangle - H_{m,\varepsilon_{n(k)}}(z_1, z_2)\}\\ &\quad + \varepsilon_{n(k)} \log\{\mu_{0,m,\varepsilon_{n(k)}}(U_r(x))\mu_{1,m,\varepsilon_{n(k)}}(U_r(y))\}. \end{aligned}$$

Since $H_{m,\varepsilon_{n(k)}}$ converges to H_m as $k \to \infty$, uniformly on every compact subset of $\mathbb{R}^d \times \mathbb{R}^d$,

$$\begin{aligned} &\inf_{(z_1,z_2)\in U_r(x)\times U_r(y)} (\langle x, z_2\rangle + \langle y, z_1\rangle - H_{m,\varepsilon_{n(k)}}(z_1, z_2)) \qquad (2.164)\\ &\to \inf_{(z_1,z_2)\in U_r(x)\times U_r(y)} (\langle x, z_2\rangle + \langle y, z_1\rangle - H_m(z_1, z_2)), \quad k \to \infty,\\ &\to 2\langle x, y\rangle - H_m(x, y), \quad r \to 0,\\ &\to 2\langle x, y\rangle - H(x, y), \quad m \to \infty. \end{aligned}$$

For sufficiently large $m \geq 1$,

$$\liminf_{\varepsilon\to 0}\{\mu_{0,m,\varepsilon}(U_r(x))\mu_{1,m,\varepsilon}(U_r(y))\} > 0. \qquad (2.165)$$

Indeed,

$$\begin{aligned} &\mu_{0,m,\varepsilon}(U_r(x))\mu_{1,m,\varepsilon}(U_r(y))\\ &= \{P_0(U_r(x)) - \mu_\varepsilon(U_r(x) \times U_m(o)^c)\}\{P_{1,\varepsilon}(U_r(y)) - \mu_\varepsilon(U_m(o)^c \times U_r(y))\}, \end{aligned}$$

$$\mu_\varepsilon(U_r(x) \times U_m(o)^c) \leq \frac{1}{m^2}\int_{\mathbb{R}^d} |z|^2 P_{1,\varepsilon}(dz) \leq \frac{2(\varepsilon d + \int_{\mathbb{R}^d} |z|^2 P_1(dz))}{m^2},$$

$$\mu_\varepsilon(U_m(o)^c \times U_r(y)) \leq \frac{1}{m^2}\int_{\mathbb{R}^d} |z|^2 P_0(dz),$$

$$\liminf_{\varepsilon\to 0} P_{1,\varepsilon}(U_r(y)) \geq P_1(U_r(y))$$

since $P_{1,\varepsilon}$ weakly converges to P_1 as $\varepsilon \to 0$. Hence (2.165) holds since for $(x, y) \in \operatorname{supp}(P_0) \times \operatorname{supp}(P_1)$,

$$(P_0 \times P_1)(U_r(x) \times U_r(y)) > 0.$$

Equations (2.163)–(2.165) implies (2.160).

The set S is cyclically monotone. Indeed, for any $k, \ell \geq 1$, $(x_1, y_1), \cdots, (x_\ell, y_\ell) \in S$, and $m \geq m_0$, from (2.153),

$$\sum_{i=1}^{\ell}(H_{m,\varepsilon_{n(k)}}(x_{i+1}, y_i) - H_{m,\varepsilon_{n(k)}}(x_i, y_i)) = 0, \tag{2.166}$$

where $x_{\ell+1} := x_1$. Let $k \to \infty$ and then $m \to \infty$. Then from (2.160),

$$\sum_{i=1}^{\ell}\langle x_{i+1} - x_i, y_i\rangle \leq \sum_{i=1}^{\ell}(H(x_{i+1}, y_i) - H(x_i, y_i)) = 0. \tag{2.167}$$

Here notice that $H(x_i, y_i)$ is finite for all $i = 1, \cdots, \ell$. □

Remark 2.14 H is lower semicontinuous since $H_m \uparrow H$ as $m \to \infty$ and since H_m is a finite convex function and hence is continuous for sufficiently large m (see the proof of Lemma 2.9). If $H(x, y)$ and $H(a, b)$ are finite, then $H(x, b)$ and $H(a, y)$ are also finite since for sufficiently large $m \geq 1$, from (2.159) and (2.166),

$$-\infty < H_m(x, b) + H_m(a, y) \leq H(x, b) + H(a, y) = H(x, y) + H(a, b) < \infty.$$

In particular,

$$H(x, y) = H(x, b) + H(a, y) - H(a, b). \tag{2.168}$$

Proof of Theorem 2.8 $\{\mu_\varepsilon\}_{\varepsilon\in(0,1]}$ is tight from (2.162) (see, e.g. [75, p. 7]). Take a weakly convergent subsequence $\{\mu_{\varepsilon_n}\}_{n\geq 1}$ and denote by μ its weak limit, where $\varepsilon_n \to 0$ as $n \to \infty$.

By taking $m_0 \geq 1$ and a subsequence $\{\varepsilon_{n(k)}\}_{k\geq 1}$, construct a convex function H as in Lemma 2.9.

From (2.160)–(2.161), we only have to show the following to complete the proof:

$$\mu(\{(x, y) \in \mathbb{R}^d \times \mathbb{R}^d | \langle x, y\rangle - H(x, y) < 0\}) = 0. \tag{2.169}$$

By the monotone convergence theorem and Lemma 2.9,

$$\begin{aligned} &\mu(\{(x, y) \in \mathbb{R}^d \times \mathbb{R}^d | \langle x, y\rangle - H(x, y) < 0\}) \\ &= \lim_{r\downarrow 0}\left\{\lim_{m\uparrow\infty} \mu(\{(x, y) \in \mathbb{R}^d \times \mathbb{R}^d | \langle x, y\rangle - H_m(x, y) < -r\})\right\}. \end{aligned} \tag{2.170}$$

For any $m \ge m_0$, $H_{m,\varepsilon_{n(k)}}$ converges to H_m as $k \to \infty$, uniformly on every compact subset of $\mathbb{R}^d \times \mathbb{R}^d$. Therefore for any $R > 0$,

$$\begin{aligned}
&\mu(\{(x,y)|\langle x,y\rangle - H_m(x,y) < -r, |x|, |y| < R\}) \qquad (2.171)\\
&\le \liminf_{k\to\infty} \mu_{\varepsilon_{n(k)}}(\{(x,y)|\langle x,y\rangle - H_m(x,y) < -r, |x|, |y| < R\})\\
&\le \liminf_{k\to\infty} \mu_{\varepsilon_{n(k)}}\left(\left\{(x,y)\middle|\langle x,y\rangle - H_{m,\varepsilon_{n(k)}}(x,y) < -\frac{r}{2}, |x|, |y| < R\right\}\right)\\
&\le \liminf_{k\to\infty} \exp\left(-\frac{r}{2\varepsilon_{n(k)}}\right) = 0 \quad \text{(from (2.154))}.
\end{aligned}$$

Notice that the set $\{(x,y)|\langle x,y\rangle - H_m(x,y) < -r, |x|, |y| < R\}$ is open since $H_m \in C(\mathbb{R}^d \times \mathbb{R}^d)$ from Lemma 2.8, (ii).

Letting $R \to \infty$ in (2.171), we obtain (2.169) from (2.170). □

Next, we prove Theorem 2.9.

Proof of Theorem 2.9 The proof of (2.140) is divided into the following:

$$\liminf_{\varepsilon\to 0} 2\varepsilon V_{S,\varepsilon}(P_0, P_{1,\varepsilon}) \ge T_M(P_0, P_1), \qquad (2.172)$$

$$\limsup_{\varepsilon\to 0} 2\varepsilon V_{S,\varepsilon}(P_0, P_{1,\varepsilon}) \le T_M(P_0, P_1) < \infty. \qquad (2.173)$$

To prove (2.172), we only have to show that for any $\{\varepsilon_n\}_{n\ge 1}$ for which $\varepsilon_n \to 0$ and $E\left[\int_0^1 |b_{\varepsilon_n}(s, X_{\varepsilon_n}(s))|^2 ds\right]$ is convergent as $n \to \infty$,

$$\lim_{n\to\infty} E\left[\int_0^1 |b_{\varepsilon_n}(s, X_{\varepsilon_n}(s))|^2 ds\right] \ge T_M(P_0, P_1) \qquad (2.174)$$

(see (2.132) and (2.139) for notation). Equation (2.174) holds since $\{X_{\varepsilon_n}(\cdot)\}_{n\ge 1}$ is tight in $C([0,1]; \mathbb{R}^d)$, and since any weak limit point $X(\cdot)$ of $\{X_{\varepsilon_n}(\cdot)\}_{n\ge 1}$ is an absolutely continuous stochastic process (see, e.g. [78, 159] or Lemmas 2–3 in [107]), and

$$\begin{aligned}
&\lim_{n\to\infty} E\left[\int_0^1 |b_{\varepsilon_n}(s, X_{\varepsilon_n}(s))|^2 ds\right] \qquad (2.175)\\
&\ge E\left[\int_0^1 \left|\frac{dX(s)}{ds}\right|^2 ds\right] \ge E[|X(1) - X(0)|^2] \ge T_M(P_0, P_1)
\end{aligned}$$

from (2.137) (see [76], p. 78 in [64], the proof of (3.17) in [107] or Lemma 3.1 in [119]).

Next, we prove (2.173). Take φ for which $P_0(D\varphi)^{-1} = P_1$, which is possible from Corollary 2.6. Then from assumption,

$$T_M(P_0, P_1) \le E[|D\varphi(X_0) - X_0|^2] \le 2\int_{\mathbb{R}^d} |x|^2(P_0(dx) + P_1(dx)) < \infty. \tag{2.176}$$

$$X_{\varepsilon,\varphi}(t) := X_0 + t(D\varphi(X_0) - X_0) + \sqrt{\varepsilon}W(t). \tag{2.177}$$

Then $P^{X_{\varepsilon,\varphi}(1)} = P_{1,\varepsilon}$, which implies (2.173).

By (2.137), (2.140), and (2.175), $D\varphi$ in Corollary 2.6 is a minimizer of $T_M(P_0, P_1)$ with $c = |y - x|^2$. The uniqueness of the minimizer of $T_M(P_0, P_1)$ with $c = |y - x|^2$ can be shown easily (see, e.g. [152, p. 69]).

Equations (2.141)–(2.142) is an easy consequence of (2.140). For $t \in [0, 1]$,

$$\begin{aligned} &|X_\varepsilon(t) - \{X_0 + t(D\varphi(X_0) - X_0)\}| \\ &\le \int_0^1 |b_\varepsilon(s, X_\varepsilon(s)) - (D\varphi(X_0) - X_0)|ds + \sqrt{\varepsilon} \sup_{0\le t\le 1} |W_{X_\varepsilon}(t)|. \end{aligned} \tag{2.178}$$

$$E\left[\sup_{0\le t\le 1} |W_{X_\varepsilon}(t)|^2\right] \le 4d \tag{2.179}$$

by Doob's inequality (see, e.g. [75, p. 34]), and from (2.140),

$$\begin{aligned} &E\left[\int_0^1 |b_\varepsilon(s, X_\varepsilon(s)) - (D\varphi(X_0) - X_0)|^2 ds\right] \\ &= E\left[\int_0^1 |b_\varepsilon(s, X_\varepsilon(s))|^2 ds + |D\varphi(X_0) - X_0|^2\right] \\ &\quad - 2E[\langle X_\varepsilon(1) - X_0 - \sqrt{\varepsilon}W_{X_\varepsilon}(1), D\varphi(X_0) - X_0\rangle] \\ &\to 2T_M(P_0, P_1) - 2E[\langle D\varphi(X_0) - X_0, D\varphi(X_0) - X_0\rangle] = 0, \quad \varepsilon \to 0. \end{aligned} \tag{2.180}$$

Indeed,

$$E[\langle W_{X_\varepsilon}(1), D\varphi(X_0) - X_0\rangle] = \langle E[W_{X_\varepsilon}(1)], E[D\varphi(X_0) - X_0]\rangle = 0.$$

For any $R > 0$, taking $f_R \in C(\mathbb{R}^d; [0, 1])$ for which $f_R(x) = 1$ $(|x| \le R)$ and $f_R(x) = 0$ $(|x| \ge R + 1)$,

$$\begin{aligned} &E[\langle X_\varepsilon(1), D\varphi(X_0) - X_0\rangle] \\ &= E[\langle X_\varepsilon(1), D\varphi(X_0) - X_0\rangle\{1 - f_R(X_\varepsilon(1)) f_R(D\varphi(X_0) - X_0)\}] \\ &\quad + E[\langle X_\varepsilon(1), D\varphi(X_\varepsilon(0)) - X_\varepsilon(0)\rangle f_R(X_\varepsilon(1)) f_R(D\varphi(X_\varepsilon(0)) - X_\varepsilon(0))]. \end{aligned}$$

$$E[|\langle X_\varepsilon(1), D\varphi(X_0) - X_0\rangle|\{1 - f_R(X_\varepsilon(1)) f_R(D\varphi(X_0) - X_0)\}]$$
$$\leq (E[|D\varphi(X_0) - X_0|^2] E[|X_\varepsilon(1)|^2; |X_\varepsilon(1)| \geq R])^{1/2}$$
$$+ (E[|X_\varepsilon(1)|^2] E[|D\varphi(X_0) - X_0|^2; |D\varphi(X_0) - X_0| \geq R])^{1/2} \to 0$$

as $R \to \infty$, uniformly in $\varepsilon \in [0,1]$. $(X_\varepsilon(0), X_\varepsilon(1))$ weakly converges to $(X_0, D\varphi(X_0))$ as $\varepsilon \to 0$ by the uniqueness of the minimizer of $T_M(P_0, P_1)$. One can assume, by taking a new probability space $(\tilde{\Omega}, \tilde{\mathbf{B}}, \tilde{P})$ if necessary, that $(X_\varepsilon(0), X_\varepsilon(1))$ converges to $(X_0, D\varphi(X_0))$ as $\varepsilon \to 0$, $\tilde{P}$–a.s., by Skhorohod's theorem (see, e.g. [75, p. 9]).

$$A := \{y \in \mathbb{R}^d | \varphi(y) < \infty, \partial\varphi(y) = \{D\varphi(y)\}\},$$

where $\partial\varphi(y)$ denotes the subdifferential of φ at y:

$$\partial\varphi(y) := \{p \in \mathbb{R}^d | \varphi(z) \geq \varphi(y) + \langle p, z - y\rangle, z \in \mathbb{R}^d\}.$$

Then $X_0 \in A$ a.s. since $P_0(dx) \ll dx$ and $\cap_{r>0}\partial\varphi(U_r(x)) = \{D\varphi(x)\}$ for any $x \in A$ (see [152, p. 54]), from which the following holds:

$$E[\langle X_\varepsilon(1), D\varphi(X_\varepsilon(0)) - X_\varepsilon(0)\rangle f_R(X_\varepsilon(1)) f_R(D\varphi(X_\varepsilon(0)) - X_\varepsilon(0))]$$
$$\to \tilde{E}[\langle D\varphi(X_0), D\varphi(X_0) - X_0\rangle f_R(D\varphi(X_0)) f_R(D\varphi(X_0) - X_0) : X_0 \in A], \quad \varepsilon \to 0,$$
$$\to E[\langle D\varphi(X_0), D\varphi(X_0) - X_0\rangle], \quad R \to \infty.$$

Equations (2.178)–(2.180) imply (2.141)–(2.142). □

We prove Corollary 2.7.

Proof of Corollary 2.7 First, we prove that $\{b_\varepsilon(t, X_\varepsilon(t))\}_{0 \leq t < 1}$ is an $(\mathscr{F}_t^{X_\varepsilon})$-martingale. The following holds from (3.7) in Sect. 3.1.1: for $t \in [0, 1)$,

$$b_\varepsilon(t, X_\varepsilon(t)) = \varepsilon D_x \log h_\varepsilon(t, X_\varepsilon(t)) = E\left[\frac{X_\varepsilon(1) - X_\varepsilon(t)}{1-t}\middle|(t, X_\varepsilon(t))\right] \tag{2.181}$$
$$= E\left[\frac{X_\varepsilon(1) - X_\varepsilon(t)}{1-t}\middle|\mathscr{F}_t^{X_\varepsilon}\right]$$

since $X_\varepsilon(t)$ is Markovian. If $0 \leq s \leq t < 1$, then

$$E[b_\varepsilon(t, X_\varepsilon(t))|\mathscr{F}_s^{X_\varepsilon}] = E\left[\frac{X_\varepsilon(1) - X_\varepsilon(t)}{1-t}\middle|\mathscr{F}_s^{X_\varepsilon}\right] \tag{2.182}$$
$$= E\left[\frac{X_\varepsilon(1) - X_\varepsilon(s)}{1-s}\middle|\mathscr{F}_s^{X_\varepsilon}\right].$$

Indeed, from the SDE for X_ε in (2.132) and (2.181),

$$E\left[\frac{X_\varepsilon(1)-X_\varepsilon(t)}{1-t}\middle|\mathscr{F}_s^{X_\varepsilon}\right]=\frac{1}{1-t}\int_t^1 E\left[b_\varepsilon(\alpha,X_\varepsilon(\alpha))\middle|\mathscr{F}_s^{X_\varepsilon}\right]d\alpha$$
$$=\frac{1}{1-t}\int_t^1 E\left[\frac{X_\varepsilon(1)-X_\varepsilon(\alpha)}{1-\alpha}\middle|\mathscr{F}_s^{X_\varepsilon}\right]d\alpha.$$

If $f(t)=\frac{1}{1-t}\int_t^1 f(\alpha)d\alpha$, $s\le t<1$, then $f\in C([s,1))$ and $f(t)=f(s)$, $s\le t<1$ since

$$\frac{d}{dt}f(t)=\frac{1}{(1-t)^2}\int_t^1 f(\alpha)d\alpha-\frac{f(t)}{1-t}=0,\quad s\le t<1.$$

Equations (2.181)–(2.182) imply that $\{b_\varepsilon(t,X_\varepsilon(t))\}_{0\le t<1}$ is an $(\mathscr{F}_t^{X_\varepsilon})$-martingale.

If $0<t<1$, then the following holds: by Doob's inequality,

$$E[\sup_{0\le s\le t}|b_\varepsilon(s,X_\varepsilon(s))-(D\varphi(X_0)-X_0)|^2] \tag{2.183}$$
$$\le 4E[|b_\varepsilon(t,X_\varepsilon(t))-(D\varphi(X_0)-X_0)|^2]$$
$$=\frac{4}{1-t}E[\int_t^1|E[b_\varepsilon(\alpha,X_\varepsilon(\alpha))-(D\varphi(X_0)-X_0)|\mathscr{F}_t^{X_\varepsilon}]|^2d\alpha]$$
$$\le\frac{4}{1-t}E[\int_t^1|b_\varepsilon(\alpha,X_\varepsilon(\alpha))-(D\varphi(X_0)-X_0)|^2d\alpha]\to 0\quad\text{(from (2.141))},$$

which completes the proof. □

2.3.2 Duality Theorem for OT by the Zero-Noise Limit of SOT

In this section, we prove the Duality Theorem for the OT by the zero-noise limit of the Duality Theorem for the SOT.

We assume that $a=\varepsilon\times Id$ and write $V(P_0,P_1)=V_\varepsilon(P_0,P_1)$ and denote by $\mathscr{T}(P_0,P_1)$ and $\mathscr{V}_\varepsilon(P_0,P_1)$ the right-hand side of (2.5) and (2.38), respectively.

If $L=L(u)$ and (A.1)–(A.2) hold, then (A.3) also holds and $V_\varepsilon(P_0,P_1)=\mathscr{V}_\varepsilon(P_0,P_1)$ from Theorem 2.4.

The following generalizes [120], where we assumed $L\in C^1(\mathbb{R}^d;[0,\infty))$ and is strictly convex.

Theorem 2.10 *Suppose that* $L = L(u)$, $c(x, y) = L(y - x)$, *and (A.1)–(A.2) hold. Then for any* $P_0, P_1 \in \mathscr{P}(\mathbb{R}^d)$,

$$\mathscr{T}(P_0, P_1) \le T(P_0, P_1) \le \liminf_{\varepsilon \to 0} V_\varepsilon(P_0, P_{1,\varepsilon}), \tag{2.184}$$

$$\mathscr{V}_\varepsilon(P_0, P_{1,\varepsilon}) \le \mathscr{T}(P_0, P_1), \quad \varepsilon > 0. \tag{2.185}$$

In particular, the following holds:

$$T(P_0, P_1) = \mathscr{T}(P_0, P_1). \tag{2.186}$$

Proof of Corollary 2.7 The first inequality in (2.184) is trivial. The second inequality in (2.184) can be proved by the lower semicontinuity of T since $P_{1,\varepsilon} \to P_1$ as $\varepsilon \to 0$ weakly. Indeed, for any $X \in \mathscr{A}$, by Jensen's inequality,

$$\begin{aligned}\int_0^1 L(\beta_X(t, X))dt &\ge L\left(\int_0^1 \beta_X(t, X)dt\right) = L(X(1) - X(0) - \sqrt{\varepsilon}W_X(1)) \\ &\ge T(g(\varepsilon, \cdot) * P^{X(0)}, P^{X(1)})\end{aligned}$$

(see (2.131) for notation). Next, we prove (2.185). For $f \in C_b^\infty(\mathbb{R}^d)$,

$$\int_{\mathbb{R}^d} f(y)P_{1,\varepsilon}(dy) - \int_{\mathbb{R}^d} \varphi(0, x; f)P_0(dx) \le \mathscr{T}(P_0, P_1). \tag{2.187}$$

Indeed,

$$\int_{\mathbb{R}^d} f(y)P_{1,\varepsilon}(dy) = \int_{\mathbb{R}^d} E[f(y + \sqrt{\varepsilon}W(1))]P_1(dy).$$

$E[f(\cdot + \sqrt{\varepsilon}W(1))] \in C_b(\mathbb{R}^d)$. $\varphi(0, \cdot; f) \in C_b(\mathbb{R}^d)$ from Lemma 2.5.

$$X^\varepsilon_{x,y}(t) := x + t(y - x) + \sqrt{\varepsilon}W(t), \quad x, y \in \mathbb{R}^d, t \in [0, 1].$$

Then from Lemma 2.5 with $Q = \delta_x$ in (2.89),

$$E[f(y + \sqrt{\varepsilon}W(1))] - \varphi(0, x; f) = E[f(X^\varepsilon_{x,y}(1))] - \varphi(0, x; f) \le L(y - x), \tag{2.188}$$

which completes the proof. □

Chapter 3
Marginal Problem

Abstract The SOT is a class of marginal problems, in that it constructs a semimartingale with given marginal distributions as a minimizer of a variational problem. In this chapter, we study the upper and lower bounds, the time-reversal, the semiconcavity, and the continuity of the infimum in Schrödinger's problem which is a typical class of the SOTs. We also study the regularity of Schrödinger's Func. Eqn. with respect to two endpoint marginal distributions and a kernel function and give an interpretation, as a mean field PDE, to the coupled Fokker–Planck HJB equation for marginal distributions of an h-path process. We consider the construction of stochastic processes with given marginals from the viewpoint of the SOT. By the OT and the superposition principle that can be considered a class of marginal problems, we also consider the SOT with a nonconvex cost in the one-dimensional case.

3.1 Schrödinger's Problem

As we discussed in Sect. 1.4, Schrödinger's problem is the variational problem on relative entropies of distributions of particles with given initial and terminal distributions. In this section, we first describe Schrödinger's problem, then we study the upper and the lower bounds and the time-reversal of the infimum in the problem. We also show the semiconcavity and the continuity, in random variables with given distributions, of the infimum, though it is convex and lower semicontinuous in marginal distributions. Lastly, we study the regularity of Schrödinger's Func. Eqn. in two endpoint marginal distributions and a kernel function and give an interpretation, as a mean field PDE, of the coupled Fokker–Planck HJB equation for marginal distributions of an h-path process.

T. Mikami, *Stochastic Optimal Transportation*, SpringerBriefs in Mathematics,
https://doi.org/10.1007/978-981-16-1754-6_3

3.1.1 Schrödinger's Problem on $\mathscr{P}(\mathbb{R}^d)$ as SOT

In this section, we consider Schrödinger's problem in the case where a kernel function q in (1.40) is the transition probability density of a diffusion process and show that Schrödinger's problem is the SOT considered in Corollary 2.3.

We first describe assumptions and then state Jamison's results.

(A.6). $\sigma(t,x) = (\sigma^{ij}(t,x))_{i,j=1}^d$, $(t,x) \in [0,1] \times \mathbb{R}^d$, is a $d \times d$-matrix. $a(t,x) := \sigma(t,x)\sigma(t,x)^t$, $(t,x) \in [0,1] \times \mathbb{R}^d$, is uniformly positive definite, bounded, once continuously differentiable, and uniformly Hölder continuous. $D_x a(t,x)$ is bounded and the first derivatives of $a(t,x)$ are uniformly Hölder continuous in x uniformly in $t \in [0,1]$.

(A.7). $\xi(t,x) : [0,1] \times \mathbb{R}^d \to \mathbb{R}^d$ is bounded, continuous, and uniformly Hölder continuous in x uniformly in $t \in [0,1]$.

Theorem 3.1 (See [80, p. 330]) *Suppose that (A.6)–(A.7) hold. Then for any $P_0 \in \mathscr{P}(\mathbb{R}^d)$, the following SDE has a unique weak solution with a positive continuous transition probability density $p(t,x;s,y)$, $0 \le t < s \le 1$, $x, y \in \mathbb{R}^d$:*

$$\begin{aligned} d\mathbf{X}(t) &= \xi(t,\mathbf{X}(t))dt + \sigma(t,\mathbf{X}(t))dW_{\mathbf{X}}(t), \quad 0 < t < 1, \\ P^{\mathbf{X}(0)} &= P_0 \end{aligned} \tag{3.1}$$

(see Sect. 2.2 for notation).

We recall Schrödinger's Func. Eqn. in our setting.

Definition 3.1 (Schrödinger's Func. Eqn.) Suppose that (A.6)–(A.7) hold. For $P_0, P_1 \in \mathscr{P}(\mathbb{R}^d)$, find a product measure $\nu_1(dx_1)\nu_2(dx_2)$ of nonnegative σ-finite Borel measures on $\mathbb{R}^d$ for which the following holds:

$$\begin{cases} P_0(dx_1) = \nu_1(dx_1) \displaystyle\int_{\mathbb{R}^d} p(0,x_1;1,x_2)\nu_2(dx_2), \\ \\ P_1(dx_2) = \nu_2(dx_2) \displaystyle\int_{\mathbb{R}^d} p(0,x_1;1,x_2)\nu_1(dx_1). \end{cases} \tag{3.2}$$

Since $p(0,x_1;1,x_2)$ is positive under (A.6)–(A.7), Schrödinger's Func. Eqn. has the unique solution for any $P_0, P_1 \in \mathscr{P}(\mathbb{R}^d)$ (see [79] and also [36]). Besides, the following is known.

Theorem 3.2 (See [80, p. 330]) *Suppose that (A.6)–(A.7) hold. Then for any $P_0, P_1 \in \mathscr{P}(\mathbb{R}^d)$ and ν_2 in (3.2),*

$$h(t,x) := \int_{\mathbb{R}^d} p(t,x;1,x_2)\nu_2(dx_2) \in C^{1,2}([0,1) \times \mathbb{R}^d), \tag{3.3}$$

$$(\partial_t + A_t)h(t,x) = 0, \quad (t,x) \in [0,1) \times \mathbb{R}^d. \tag{3.4}$$

Here

$$A_t := \frac{1}{2}Trace(a(t,x)D_x^2) + \langle \xi(t,x), D_x \rangle.$$

Theorem 3.3 (Markovian Reciprocal Process (See [80, Theorem 2])) *Suppose that (A.6)–(A.7) hold. Then for any $P_0 \in \mathscr{P}(\mathbb{R}^d)$ and $P_1 \in \mathscr{P}_{ac}(\mathbb{R}^d)$, there exists a unique weak solution to the following SDE:*

$$\begin{aligned} dX(t) &= \{a(t, X(t))D_x \log h(t, X(t)) + \xi(t, X(t))\}dt \\ &\quad + \sigma(t, X(t))dW_X(t), \quad 0 < t < 1, \end{aligned} \tag{3.5}$$

$$P^{X(t)} = P_t, \quad t = 0, 1.$$

Besides,

$$P^{((X(0),X(1))}(dxdy) = \nu_1(dx)p(0,x;1,y)\nu_2(dy), \tag{3.6}$$

$$P^{X(t)}(dx) = \left(\int_{\mathbb{R}^d} \nu_1(dx_1)p(0,x_1;t,x)\right)h(t,x)dx, \quad 0 \le t \le 1, \tag{3.7}$$

where

$$\int_{\mathbb{R}^d} \nu_1(dx_1)p(0,x_1;0,x)dx := \nu_1(dx),$$

$$h(1,x) = \int_{\mathbb{R}^d} \nu_2(dx_2)p(1,x;1,x_2) := \frac{\nu_2(dx)}{dx}.$$

Schrödinger's Func. Eqn. is Euler's equation for Schrödinger's problem (see (1.40)). Besides, Schrödinger's problem is the SOT $V_S(P_0, P_1) := V(P_0, P_1)$ in the case where

$$L = \frac{1}{2}|\sigma^{-1}(t,x)(u - \xi(t,x))|^2.$$

Proposition 3.1 *Suppose that (A.6)–(A.7) hold. Then any $P_0 \in \mathscr{P}(\mathbb{R}^d)$ and $P_1 \in \mathscr{P}_{ac}(\mathbb{R}^d)$,*

$$\inf\{H(\mu(dxdy)|P_0(dx)p(0,x;1,y)dy)|\mu_1 = P_0, \mu_2 = P_1\} = V_S(P_0, P_1) \tag{3.8}$$

(see (1.39) and (2.2) for notation). $\{X(t)\}_{0 \le t \le 1}$ in (3.5) is the unique minimizer, provided (3.8) is finite.

We explain why this is true. For $\{X(t)\}_{0\le t\le 1} \in \mathscr{A}$ (see Sect. 2.2 for notation) such that $P^{X(t)} = P_t, t = 0, 1$, by Jensen's inequality,

$$E\left[\int_0^1 \frac{1}{2}|\sigma^{-1}(t, X(t))(\beta_X(t, X) - \xi(t, X(t)))|^2 dt\right] \tag{3.9}$$
$$= H(P^{X(\cdot)}|P^{\mathbf{X}(\cdot)})$$
$$\ge H(P^{(X(0),X(1))}|P^{(\mathbf{X}(0),\mathbf{X}(1))}) \ge \text{l.h.s. of (3.8)}.$$

Theorem 3 in [135] implies that $\nu_1(dx)p(0, x; 1, y)\nu_2(dy)$ is the unique minimizer of the l.h.s. of (3.8), provided the l.h.s. of (3.8) is finite since

$$P_0(dx)p(0, x; 1, y)dy = \frac{p(0, x; 1, y)}{p(1, y)} P_0(dx)p(1, y)dy \ll P_0(dx)p(1, y)dy$$

(see (C.1) in Sect. 2.2.3 for notation). This implies that $\{X(t)\}_{0\le t\le 1}$ in (3.5) is the unique minimizer of $V_S(P_0, P_1)$ and that (3.8) holds, provided the l.h.s. of (3.8) is finite (see Corollary 2.3). Indeed, for $\{X(t)\}_{0\le t\le 1}$ in (3.5),

$$\text{l.h.s. of (3.8)} = H(P^{(X(0),X(1))}|P^{(\mathbf{X}(0),\mathbf{X}(1))}) = H(P^{X(\cdot)}|P^{\mathbf{X}(\cdot)}).$$

Remark 3.1 If $H(P^{X(\cdot)}|P^{\mathbf{X}(\cdot)})$ is finite, then $P^{X(\cdot)} \ll P^{\mathbf{X}(\cdot)}$. In particular, $P^{X(1)} \in \mathscr{P}_{ac}(\mathbb{R}^d)$ since $P^{\mathbf{X}(1)} \in \mathscr{P}_{ac}(\mathbb{R}^d)$, provided (A.6)–(A.7) hold.

3.1.2 Bounds of Schrödinger's Problem

In this section, we give the upper and lower bounds of $V_S(P_0, P_1)$ in (3.8) by Schrödinger's Func. Eqn. and compare the upper bound with that obtained in Sect. 2.2.3.

Let λ_1 and λ_d denote the infimum of the least eigenvalue and the supremum of the largest eigenvalue of $a(t, x)$, respectively.

$$||f||_\infty := \sup\{|f(t, x)||(t, x) \in [0, 1] \times \mathbb{R}^d\}, \quad f \in C_b([0, 1] \times \mathbb{R}^d).$$

From Corollary 2.3 and Lemma 2.7 in Sect. 2.2, the following holds.

Proposition 3.2 *Suppose that (A.4)' in Sect. 2.2.1 holds. Then for any* $P_0, P_1 \in \mathscr{P}(\mathbb{R}^d)$ *for which (C.1) in Sect. 2.2.3 holds,*

$$\begin{aligned} &V_S(P_0, P_1) \\ &\le \frac{3}{\lambda_1}\int_{\mathbb{R}^d\times\mathbb{R}^d} |y-x|^2 p(0,x)p(1,y)dxdy + \frac{5}{4\lambda_1}(\|D_x^t a\|_\infty^2 + \|\xi\|_\infty^2) \\ &\quad + \frac{\lambda_d}{2}\int_{\mathbb{R}^d}(p(0,x)|D_x \log p(0,x)|^2 + p(1,x)|D_x\log p(1,x)|^2)dx. \end{aligned} \tag{3.10}$$

Proof We only have to consider the case where (C.2)' with $r=2$ holds. Otherwise, the right-hand side of (3.10) is infinite. From Corollary 2.3 and Lemma 2.7,

$$\begin{aligned} &V_S(P_0, P_1) \\ &\le \frac{3}{4\lambda_1}\int_0^1 dt\int_{\mathbb{R}^d\times\mathbb{R}^d}\{|2(y-x)|^2 + \|D_x^t a\|_\infty^2 + \|\xi\|_\infty^2\}p(0,x)p(1,y)dxdy \\ &\quad + \frac{1}{2}\int_0^1 \min\Bigg(\int_{\mathbb{R}^d}\left\{\lambda_d\left|\frac{1}{1-t}D_x\log p(0,x)\right|^2 + \frac{\|\xi\|_\infty^2}{\lambda_1}\right\}p(0,x)dx, \\ &\qquad\qquad \int_{\mathbb{R}^d}\left\{\lambda_d\left|\frac{1}{t}D_y\log p(1,y)\right|^2 + \frac{\|\xi\|_\infty^2}{\lambda_1}\right\}p(1,y)dy\Bigg)dt \\ &= \frac{3}{\lambda_1}\int_{\mathbb{R}^d\times\mathbb{R}^d}|y-x|^2 p(0,x)p(1,y)dxdy + \frac{1}{4\lambda_1}(3\|D_x^t a\|_\infty^2 + 5\|\xi\|_\infty^2) \\ &\quad + \frac{\lambda_d}{2}\int_0^1 \min\Bigg(\int_{\mathbb{R}^d}\left|\frac{1}{1-t}D_x\log p(0,x)\right|^2 p(0,x)dx, \\ &\qquad\qquad \int_{\mathbb{R}^d}\left|\frac{1}{t}D_y\log p(1,y)\right|^2 p(1,y)dy\Bigg)dt. \end{aligned} \tag{3.11}$$

The last integral on the right-hand side of (3.11) can be estimated as follows:

$$\begin{aligned} &\int_0^1 \min\Bigg(\int_{\mathbb{R}^d}\left|\frac{1}{1-t}D_x\log p(0,x)\right|^2 p(0,x)dx, \\ &\qquad \int_{\mathbb{R}^d}\left|\frac{1}{t}D_x\log p(1,x)\right|^2 p(1,x)dx\Bigg)dt \end{aligned} \tag{3.12}$$

$$\le \int_0^{\frac{1}{2}}\frac{1}{(1-t)^2}dt\int_{\mathbb{R}^d}|D_x\log p(0,x)|^2 p(0,x)dx$$

$$+\int_{\frac{1}{2}}^{1}\frac{1}{t^2}dt\int_{\mathbb{R}^d}|D_x\log p(1,x)|^2p(1,x)dx$$

$$=\int_{\mathbb{R}^d}(|D_x\log p(0,x)|^2p(0,x)+|D_x\log p(1,x)|^2p(1,x))dx.$$

Equations (3.11)–(3.12) complete the proof.

In Theorem 3.3,

$$\varphi_h(t,x):=\log h(t,x),\quad (t,x)\in[0,1]\times\mathbb{R}^d \tag{3.13}$$

and let $q(s,x;t,y)$, $0\le s<t\le 1$, $x,y\in\mathbb{R}^d$ denote the transition probability density of $\{X(t)\}_{0\le t\le 1}$ in (3.5). Then the following is known:

$$q(s,x;t,y)=\exp(\varphi_h(t,y)-\varphi_h(s,x))p(s,x;t,y), \tag{3.14}$$

$$V_S(P_0,P_1)=\int_{\mathbb{R}^d}\varphi_h(1,y)p(1,y)dy-\int_{\mathbb{R}^d}\varphi_h(0,x)P_0(dx),$$

provided $V_S(P_0,P_1)$ is finite (see [45, 62, 119, 135, 156] and also Corollary 2.3 in Sect. 2.2.1). Here $\log 0:=-\infty$ and $e^{-\infty}:=0$.

We have the following from Schrödinger's Func. Eqn. (see (2.145) for notation).

Proposition 3.3 *Suppose that (A.6)–(A.7) hold. Then for any* $P_0, P_1\in\mathscr{P}(\mathbb{R}^d)$ *for which* $\mathscr{S}(P_1)$ *is finite (see (2.145) for notation),*

$$\begin{aligned}&\mathscr{S}(P_1)-\int_{\mathbb{R}^d\times\mathbb{R}^d}\{\log p(0,x;1,y)\}P_0(dx)q(0,x;1,y)dy \\ &\le V_S(P_0,P_1)\\ &\le \mathscr{S}(P_1)-\int_{\mathbb{R}^d\times\mathbb{R}^d}\{\log p(0,x;1,y)\}P_0(dx)P_1(dy).\end{aligned} \tag{3.15}$$

Proof We prove the first inequality in (3.15). From Proposition 3.1 and (3.14), by Jensen's inequality,

$$\begin{aligned}&V_S(P_0,P_1)\\ &=\int_{\mathbb{R}^d}dy\int_{\mathbb{R}^d}(\log q(0,x;1,y)-\log p(0,x;1,y))P_0(dx)q(0,x;1,y)\\ &\ge\mathscr{S}(P_1)-\int_{\mathbb{R}^d\times\mathbb{R}^d}\{\log p(0,x;1,y)\}P_0(dx)q(0,x;1,y)dy.\end{aligned} \tag{3.16}$$

The second inequality in (3.15) can be shown from the following: from (3.8),

$$V_S(P_0, P_1) \le H(P_0(dx)P_1(dy)|P_0(dx)p(0, x; 1, y)dy), \tag{3.17}$$

which completes the proof.

By the log-Sobolev inequality for the Gaussian measure (see, e.g. [152, p. 547, (21.5)]),

$$\int_{\mathbb{R}^d} p(1, y) \log\left(\frac{p(1, y)}{g(1, y)}\right) dy \le \frac{1}{2}\int_{\mathbb{R}^d} \left|D_y \log \frac{p(1, y)}{g(1, y)}\right|^2 p(1, y)dy, \tag{3.18}$$

where g denotes a Gaussian kernel (see (1.47) for notation). The following holds under (A.6)–(A.7) (see [6, 66]): there exist constants $C_1, C_2 > 0$ such that

$$-C_1 + C_2^{-1}|x - y|^2 \le -\log p(0, x; 1, y) \le C_1 + C_2|x - y|^2. \tag{3.19}$$

The following implies that Schrödinger's Func. Eqn. gives a better upper bound of $V_S(P_0, P_1)$ than our approach by the Duality Theorem for $V_S(P_0, P_1)$, in that the finiteness of $\int_{\mathbb{R}^d} |D_x \log p(0, x)|^2 p(0, x)dx$ is not assumed.

From the second inequality in (3.15) and (3.18), the following holds.

Corollary 3.1 *Suppose that (A.6)–(A.7) hold. Then for any $P_0 \in \mathscr{P}_2(\mathbb{R}^d)$ and $P_1 \in \mathscr{P}_{2,ac}(\mathbb{R}^d)$ for which $\int_{\mathbb{R}^d} |D_x \log p(1, x)|^2 P_1(dx)$ is finite,*

$$\begin{aligned} V_S(P_0, P_1) \le\ & \frac{1}{2}\int_{\mathbb{R}^d} (|D_x \log p(1, x)|^2 + 2\langle D_x \log p(1, x), x\rangle)p(1, x)dx \qquad (3.20)\\ & - \frac{d}{2}\log(2\pi) - \int_{\mathbb{R}^d\times\mathbb{R}^d} \{\log p(0, x; 1, y)\}P_0(dx)P_1(dy). \end{aligned}$$

From Proposition 3.3, (3.16), and (3.19), the following also holds (see Remark 3.1).

Corollary 3.2 *Suppose that (A.6)–(A.7) hold. Then for any $P_0, P_1 \in \mathscr{P}_2(\mathbb{R}^d)$, $V_S(P_0, P_1)$ is finite if and only if $\mathscr{S}(P_1)$ is finite.*

3.1.3 Time-Reversal of Schrödinger's Problem

In this section, we consider the time-reversal of Schrödinger's problem on [0, 1]. We only consider distributions of semimartingales in $C([0, 1]; \mathbb{R}^d)$.

We assume that (A.6)–(A.7) in Sect. 3.1.1 hold. Then for $\mathbf{X}$ in (3.1), $P^{\mathbf{X}(t)}$ has a density $p^{\mathbf{X}}(t, x)$ for $t > 0$ (see Theorem 3.1).

For a function f which depends on $t \in [0, 1]$,

$$\overline{f}(t, \cdot) := f(1 - t, \cdot).$$

For $X = \{X(t)\}_{0 \le t \le 1} \in \mathscr{A}$,

$$\begin{aligned} H(P^X | P^{\mathbf{X}}) &= H(P^{\overline{X}} | P^{\overline{\mathbf{X}}}) \qquad (3.21) \\ &= H(P^{X(1)} | P^{\mathbf{X}(1)}) + H(P^{\overline{X}} | P^{\overline{X}(0)}(dx) P^{\overline{\mathbf{X}}}(\cdot | \overline{\mathbf{X}}(0) = x)) \end{aligned}$$

since

$$\begin{aligned} P^{\overline{X}}(\cdot) &= P^{\overline{X}(0)}(dx) P^{\overline{X}}(\cdot | \overline{X}(0) = x), \quad P^{\overline{X}(0)} = P^{X(1)}, \\ P^{\overline{\mathbf{X}}}(\cdot) &= P^{\overline{\mathbf{X}}(0)}(dx) P^{\overline{\mathbf{X}}}(\cdot | \overline{\mathbf{X}}(0) = x), \quad P^{\overline{\mathbf{X}}(0)} = P^{\mathbf{X}(1)}. \end{aligned}$$

Suppose that

$$\xi(t, x) = \frac{1}{2} \{D_x^t a(t, x)\}^t = \frac{1}{2} \left(\sum_{i=1}^{d} \frac{\partial a_{ij}(t, x)}{\partial x_i} \right)_{j=1}^{d}. \qquad (3.22)$$

Then the following holds: for any $f \in C_b^2(\mathbb{R}^d)$

$$\begin{aligned} &\int_{\mathbb{R}^d} f(x)(p^{\mathbf{X}}(t, x) - p^{\mathbf{X}}(0, x))dx \qquad (3.23) \\ &= \int_{[0,t] \times \mathbb{R}^d} \frac{1}{2} div_x(a(s, x) Df(x)) ds p^{\mathbf{X}}(s, x) dx \end{aligned}$$

and $f \mapsto div_x(a(t, x)Df(x))$ is self-adjoint for $t \in [0, 1]$. Let $\mathbf{Y} = \{\mathbf{Y}(t)\}_{0 \le t \le 1}$ be the unique weak solution to the following SDE:

$$\begin{aligned} d\mathbf{Y}(t) &= \overline{\xi}(t, \mathbf{Y}(t))dt + \overline{\sigma}(t, \mathbf{Y}(t))dW_{\mathbf{Y}}(t), \quad 0 < t < 1, \qquad (3.24) \\ P^{\mathbf{Y}(0)} &= P^{\mathbf{X}(1)}. \end{aligned}$$

Then the transition densities of $\mathbf{Y}$ and $\overline{\mathbf{X}}$ are the following:

$$P(\mathbf{Y}(t) \in dy | \mathbf{Y}(s) = x) = p(1 - t, y; 1 - s, x)dy, \qquad (3.25)$$

$$P(\overline{\mathbf{X}}(t) \in dy | \overline{\mathbf{X}}(s) = x) = \frac{p^{\mathbf{X}}(1 - t, y)}{p^{\mathbf{X}}(1 - s, x)} p(1 - t, y; 1 - s, x)dy, \qquad (3.26)$$

$0 \le s < t \le 1, x, y \in \mathbb{R}^d$ (see Chapter 6.4 in [65] and also [66]). Indeed, since $\mathbf{X}$ is Markovian, so is $\overline{\mathbf{X}}$ and

$$\begin{aligned} P(\overline{\mathbf{X}}(s) \in dx, \overline{\mathbf{X}}(t) \in dy) &= P(\mathbf{X}(1-s) \in dx, \mathbf{X}(1-t) \in dy) \\ &= p^{\mathbf{X}}(1-t, y)dyp(1-t, y; 1-s, x)dx \\ &= p^{\mathbf{X}}(1-s, x)dx \frac{p^{\mathbf{X}}(1-t, y)}{p^{\mathbf{X}}(1-s, x)} p(1-t, y; 1-s, x)dy \end{aligned}$$

(see [79] on Bernstein processes). The following is also known (see [73] and also [122] for related topics): if a and ξ are uniformly Lipschitz continuous in x, uniformly in t and if $|p^{\mathbf{X}}(t, x)|^2$ and $\langle a(t, x)D_x p^{\mathbf{X}}(t, x), D_x p^{\mathbf{X}}(t, x)\rangle$ are locally integrable on $(0, 1] \times \mathbb{R}^d$, then on $[0, 1)$,

$$d\overline{\mathbf{X}}(t) = \left\{ \frac{\{D_x^t(\overline{a}(t, \overline{\mathbf{X}}(t))\overline{p}^{\mathbf{X}}(t, \overline{\mathbf{X}}(t)))\}^t}{\overline{p}^{\mathbf{X}}(t, \overline{\mathbf{X}}(t))} - \overline{\xi}(t, \overline{\mathbf{X}}(t)) \right\} dt + \overline{\sigma}(t, \overline{\mathbf{X}}(t)) dW_{\overline{\mathbf{X}}}(t). \tag{3.27}$$

If $\mathscr{S}(P^{X(1)})$ and $\mathscr{S}(P^{X(0)})$ is finite, then the following holds from (3.21):

$$H(P^X|P^{\mathbf{X}}) = \mathscr{S}(P^{X(1)}) - \mathscr{S}(P^{X(0)}) + H(P^{\overline{X}}|P^{\overline{X}(0)}(dx)P^{\mathbf{Y}}(\cdot|\mathbf{Y}(0) = x)), \tag{3.28}$$

since from (3.25)–(3.26),

$$P^{\overline{\mathbf{X}}}(\cdot) = \frac{p^{\mathbf{X}}(0, \mathbf{Y}(1))}{p^{\mathbf{X}}(1, \mathbf{Y}(0))} P^{\mathbf{Y}}(\cdot).$$

Let $\overline{V}_S(P_0, P_1)$ denote $V_S(P_0, P_1)$ with a and ξ replaced by $\overline{a}$ and $\overline{\xi}$. Then the following also holds from (3.28) since the assumption in (3.27) holds under that in the following proposition (see [6]).

Proposition 3.4 *Suppose that (A.6) holds and that a and $\xi = \frac{1}{2}\{D_x^t a(t, x)\}^t$ are uniformly Lipschitz continuous in x, uniformly in t. Then for any $P_0, P_1 \in \mathscr{P}(\mathbb{R}^d)$ for which $\mathscr{S}(P_0)$, $\mathscr{S}(P_1)$, and $V_S(P_0, P_1)$ are finite and for the unique minimizer X of $V_S(P_0, P_1)$, $\overline{X}$ is the unique minimizer of $\overline{V}_S(P_1, P_0)$ and the following holds.*

$$V_S(P_0, P_1) = \mathscr{S}(P_1) - \mathscr{S}(P_0) + \overline{V}_S(P_1, P_0). \tag{3.29}$$

3.1.4 Semiconcavity and Continuity of Schrödinger's Problem

The Duality Theorem for V (see Theorem 2.4 in Sect. 2.2) implies that

$$\mathscr{P}(\mathbb{R}^d) \times \mathscr{P}(\mathbb{R}^d) \ni (P, Q) \mapsto V(P, Q)$$

is convex and lower semicontinuous (see also (2.64)).

In this section, under some assumptions, we show that for a fixed $Q \in \mathscr{P}_{2,ac}(\mathbb{R}^d)$,

$$L^2(\Omega, P; \mathbb{R}^d) \ni X \mapsto V_S(P^X, Q) \tag{3.30}$$

is semiconcave and is continuous, i.e., there exists $C > 0$ such that

$$L^2(\Omega, P; \mathbb{R}^d) \ni X \mapsto V_S(P^X, Q) - CE[|X|^2] \tag{3.31}$$

is concave and is continuous. Here $L^2(\Omega, P; \mathbb{R}^d)$ denotes the space of all square integrable functions from a probability space $(\Omega, \mathscr{F}, P)$ to $(\mathbb{R}^d, \mathbf{B}(\mathbb{R}^d))$. Let W_2 denote the Wasserstein distance of order 2, i.e. $T^{1/2}$ with $c = |y - x|^2$ in (2.3):

$$\begin{aligned} W_2(P_0, P_1) &:= \left(\inf \left\{ \int_{\mathbb{R}^d \times \mathbb{R}^d} |y - x|^2 \mu(dxdy) \middle| \mu \in \mathscr{A}(P_0, P_1) \right\} \right)^{1/2} \quad (3.32) \\ &= (\inf\{E[|X_1 - X_0|^2] | P^{X_i} = P_i, i = 0, 1\})^{1/2}, \quad P_0, P_1 \in \mathscr{P}(\mathbb{R}^d). \end{aligned}$$

We also show the Lipschitz continuity of $\mathscr{P}_2(\mathbb{R}^d) \ni P \mapsto V_S(P, Q)$ in W_2.

In section 6 of [27], P. Cardaliaguet considered a functional of probability measure as a function of a random variable and discussed the notion of derivatives in random variables. The study of the regularity of (3.30) is the first step to the theory of the Hamilton–Jacobi equation for the SOT. The concavity of (3.31) is useful to prove the uniqueness of a variational problem (see Remark 3.2, (ii)).

We recall the definition of displacement convexity.

Definition 3.2 (Displacement Convexity (See [102])) Let $G : \mathscr{P}(\mathbb{R}^d) \longrightarrow \mathbb{R} \cup \{\infty\}$. G is displacement convex if the following is convex: for any $\rho_0, \rho_1 \in \mathscr{P}(\mathbb{R}^d)$ and convex function $\varphi : \mathbb{R}^d \longrightarrow \mathbb{R} \cup \{\infty\}$,

$$[0, 1] \ni t \mapsto G(\rho_t), \tag{3.33}$$

where $\rho_t := \rho_0(id + t(D\varphi - id))^{-1}, 0 < t < 1$, provided $\rho_1 = \rho_0(D\varphi)^{-1}$ and ρ_t can be defined. Here id denotes an identity mapping.

Recall that a convex function is differentiable dx—a.e. in the interior of its domain (see, e.g. [152]) and $\rho_t, 0 < t < 1$ in (3.33) is well defined if $\rho_0 \in \mathscr{P}_{ac}(\mathbb{R}^d)$. φ in (3.33) exists if $\rho_0 \in \mathscr{P}_{2,ac}(\mathbb{R}^d)$ and if $\rho_1 \in \mathscr{P}_2(\mathbb{R}^d)$ (see Theorem 2.9).

(A.4)' in Sect. 2.2.1 implies (A.6)–(A.7). The following implies that $\mathscr{P}(\mathbb{R}^d) \ni P \mapsto V_S(P, Q)$ is displacement semiconcave for a fixed $Q \in \mathscr{P}(\mathbb{R}^d)$ and will be proved later in this section.

Theorem 3.4 (See [118]) *Suppose that (A.4)' holds and that there exists a constant $C > 0$ such that $x \mapsto \log p(0, x; 1, y) + C|x|^2$ is convex for any $y \in \mathbb{R}^d$. Then for any $Q \in \mathscr{P}_{ac}(\mathbb{R}^d)$, $X_i \in L^2(\Omega, P; \mathbb{R}^d)$, $i = 1, 2$, and $\lambda_1 \in (0, 1)$,*

$$\sum_{i=1}^{2} \lambda_i V_S(P^{X_i}, Q) - C\lambda_1\lambda_2 E[|X_1 - X_2|^2] \le V_S(P^{\sum_{i=1}^{2}\lambda_i X_i}, Q), \tag{3.34}$$

where $\lambda_2 := 1 - \lambda_1$. Equivalently, the following is concave:

$$L^2(\Omega, P; \mathbb{R}^d) \ni X \mapsto V_S(P^X, Q) - CE[|X|^2]. \tag{3.35}$$

In particular, the following is displacement convex:

$$\mathscr{P}_{2,ac}(\mathbb{R}^d) \ni P \mapsto -V_S(P, Q) + C\int_{\mathbb{R}^d} |x|^2 P(dx). \tag{3.36}$$

Remark 3.2 (i) Suppose that $a_{ij} = a_{ij}(x)$, $\xi_i = \xi_i(x) \in C_b^\infty(\mathbb{R}^d)$ and that $a(x)$ is uniformly nondegenerate. Then $D_x^2 \log p(0, x; 1, y)$ is bounded (see [143, Theorem B]). In particular, there exists a constant $C > 0$ such that $x \mapsto \log p(0, x; 1, y) + C|x|^2$ is convex for any $y \in \mathbb{R}^d$. (ii) For $C > 0$ and $P, Q \in \mathscr{P}(\mathbb{R}^d)$,

$$\Psi_{Q,C}(P) := \mathscr{S}(P) - V_S(P, Q) + C\int_{\mathbb{R}^d \times \mathbb{R}^d} |x - y|^2 P(dx)Q(dy). \tag{3.37}$$

$\Psi_{Q,C}(P)$ plays a crucial role in the construction of moment measures by the SOT (see [117] and also [138] for the approach by the OT). Since $\mathscr{P}_{ac}(\mathbb{R}^d) \ni P \mapsto \mathscr{S}(P)$ is strictly displacement convex from Theorem 2.2 in [102], so is $\mathscr{P}_{2,ac}(\mathbb{R}^d) \ni P \mapsto \Psi_{Q,C}(P)$ under the assumption of Theorem 3.4.

Suppose that (A.6)–(A.7) hold. Then from Corollary 3.2, for any $Q \in \mathscr{P}_2(\mathbb{R}^d)$ such that $\mathscr{S}(Q)$ is finite, $\mathscr{P}_2(\mathbb{R}^d) \ni P \mapsto V_S(P, Q) \in [0, \infty)$. Since (A.4)' implies (A.6)–(A.7), we obtain the following in the same way as in the finite-dimensional case.

Corollary 3.3 (See [118]) *Suppose that the same assumption as in Theorem 3.4 holds. Then for any $Q \in \mathscr{P}_{2,ac}(\mathbb{R}^d)$ for which $\mathscr{S}(Q)$ is finite, the following holds: for $P_0, P_1 \in \mathscr{P}_2(\mathbb{R}^d)$,*

$$\begin{aligned} &|V_S(P_0, Q) - V_S(P_1, Q)| \\ &\le f(\max(\|x\|_{L^2(P_0)}, \|x\|_{L^2(P_1)}), \|x\|_{L^2(Q)})W_2(P_0, P_1), \end{aligned} \tag{3.38}$$

where $||x||_{L^2(P)} := (\int_{\mathbb{R}^d} |x|^2 P(dx))^{1/2}$, $P \in \mathscr{P}_2(\mathbb{R}^d)$ *and*

$$f(x, y) := 2C_2 x^2 + 2(C_2 y^2 + C_1) + C$$

(see (3.19) for notation).

In particular, if $p(0, x; 1, y) = g(a, y - x)$, $a > 0$, *then the following holds:*

$$|V_S(P_0, Q) - V_S(P_1, Q)| \tag{3.39}$$
$$\leq \frac{1}{2a}\{||x||_{L^2(P_0)} + ||x||_{L^2(P_1)} + 2(1 + \max(\sigma_0, \sigma_1))||x||_{L^2(Q)}\}W_2(P_0, P_1),$$

where

$$\sigma_i := \left(\int_{\mathbb{R}^d}\left(x - \int_{\mathbb{R}^d} y P_i(dy)\right)^2 P_i(dx)\right)^{1/2}, \quad i = 0, 1.$$

Remark 3.3 If $a(t, x) = \varepsilon > 0$ and $\xi = 0$, then $p(0, x; 1, y) = g(\varepsilon, y - x)$. In the setting of Corollary 3.3, $2\varepsilon V_S(P_i, Q) \to W_2(P_i, Q)^2$, $\varepsilon \to 0$, $i = 0, 1$ (see Proposition 2.4 in Sect. 2.3 and also [95]). Since W_2 is a metric on $\mathscr{P}_2(\mathbb{R}^d)$ (see, e.g. [152]), for any $P_0, P_1, Q \in \mathscr{P}_2(\mathbb{R}^d)$,

$$|W_2(P_0, Q)^2 - W_2(P_1, Q)^2| = (W_2(P_0, Q) + W_2(P_1, Q))|W_2(P_0, Q) - W_2(P_1, Q)|$$
$$\leq \{W_2(P_0, Q) + W_2(P_1, Q)\}W_2(P_0, P_1)$$
$$\leq \{||x||_{L^2(P_0)} + ||x||_{L^2(P_1)} + 2||x||_{L^2(Q)}\}W_2(P_0, P_1).$$

We would like to study the best Lipschitz constant of $V_S(P, Q)$ in future.

For $P, Q \in \mathscr{P}(\mathbb{R}^d)$, let $\mu(dxdy; P, Q)$ denote the joint distribution at $t = 0, 1$ of the minimizer of $V_S(P, Q)$, provided $V_S(P, Q)$ is finite. For C_2 in (3.19) which depends on the lower bound of $\log p(0, x; 1, y)$, we discuss the weak lower semicontinuity of

$$\mathscr{P}_2(\mathbb{R}^d) \ni P \mapsto -V_S(P, Q) + C_2 \int_{\mathbb{R}^d \times \mathbb{R}^d} |x - y|^2 P(dx) Q(dy)$$

and the continuity of

$$L^2(\Omega, P; \mathbb{R}^d) \ni X \mapsto V_S(P^X, Q).$$

Here the assumption is weaker than Theorem 3.4 since we do not use the Duality Theorem. The proof is given in the end of this section.

Proposition 3.5 (See [118]) *Suppose that (A.6)–(A.7) hold. For $P, Q \in \mathscr{P}_2(\mathbb{R}^d)$, if $\mathscr{S}(Q)$ is finite, then the following holds:*

$$-V_S(P,Q) = H(P \times Q|\mu(P,Q)) - \mathscr{S}(Q) + \int_{\mathbb{R}^d \times \mathbb{R}^d} \log p(0,x;1,y) P(dx)Q(dy). \tag{3.40}$$

In particular, the following is weakly lower semicontinuous:

$$\mathscr{P}_2(\mathbb{R}^d) \ni P \mapsto -V_S(P,Q) + C_2 \int_{\mathbb{R}^d \times \mathbb{R}^d} |x-y|^2 P(dx)Q(dy). \tag{3.41}$$

The following is also continuous:

$$L^2(\Omega, P; \mathbb{R}^d) \ni X \mapsto V_S(P^X, Q). \tag{3.42}$$

In particular, $\mathscr{P}_2(\mathbb{R}^d) \ni P \mapsto V_S(P,Q)$ is continuous in the topology induced by W_2.

Remark 3.4 In the same way as in the proof of Proposition 3.5, in the setting of Proposition 3.5, one can show that the following holds:

$$V_S(P,Q) = H(\mu(P,Q)|P \times Q) + \mathscr{S}(Q) - \int_{\mathbb{R}^d \times \mathbb{R}^d} \log p(0,x;1,y) \mu(dxdy; P, Q). \tag{3.43}$$

$I(X;Y) := H(P^{(X,Y)}|P^X \times P^Y)$ is called the mutual information of random variables X and Y (see, e.g. [41]). From (3.40) and (3.43),

$$\begin{aligned} &\int_{\mathbb{R}^d \times \mathbb{R}^d} \log p(0,x;1,y)\{\mu(dxdy;P,Q) - P(dx)Q(dy)\} \\ &= H(\mu(P,Q)|P \times Q) + H(P \times Q|\mu(P,Q)) \geq 0, \end{aligned} \tag{3.44}$$

which means that for an h-path process $\{X(t)\}_{0 \leq t \leq 1}$ with $p(0,x;1,y) = g(a, y-x), a > 0$, the covariance $Cov(X(0), X(1))$ is nonnegative. We would like to know the role of mutual information in Schrödinger's problem in future. Indeed, one might develop a new research field between Schrödinger's problem and information theory.

We prove Theorem 3.4, Corollary 3.3, and Proposition 3.5.

Proof of Theorem 3.4 For any $f_i \in C_b^\infty(\mathbb{R}^d)$, $\varphi_i(x) := \varphi(0, x; f_i)$ (see (2.49) with $c = 0$ for notation). Then

$$\sum_{i=1}^{2} \lambda_i \left\{ \int_{\mathbb{R}^d} f_i(x) Q(dx) - \int_{\mathbb{R}^d} \varphi_i(x) P^{X_i}(dx) \right\} \tag{3.45}$$
$$\le V_S(P^{\sum_{i=1}^{2} \lambda_i X_i}, Q) + \lambda_1 \lambda_2 C E[|X_1 - X_2|^2].$$

Indeed,

$$\sum_{i=1}^{2} \lambda_i \left\{ \int_{\mathbb{R}^d} f_i(x) Q(dx) - \int_{\mathbb{R}^d} \varphi_i(x) P^{X_i}(dx) \right\}$$
$$= \int_{\mathbb{R}^d} \sum_{i=1}^{2} \lambda_i f_i(x) Q(dx) - E\left[\sum_{i=1}^{2} \lambda_i \{\varphi_i(X_i) + C|X_i|^2\} \right] + CE\left[\sum_{i=1}^{2} \lambda_i |X_i|^2 \right],$$

$$\int_{\mathbb{R}^d} \sum_{i=1}^{2} \lambda_i f_i(x) Q(dx) \le V_S(P^{\sum_{i=1}^{2} \lambda_i X_i}, Q) + \int_{\mathbb{R}^d} \varphi\left(0, x; \sum_{i=1}^{2} \lambda_i f_i\right) P^{\sum_{i=1}^{2} \lambda_i X_i}(dx),$$

by the Duality Theorem for $V_S(P^{\sum_{i=1}^{2} \lambda_i X_i}, Q)$ (see Corollary 2.3).

$$\int_{\mathbb{R}^d} \varphi\left(0, x; \sum_{i=1}^{2} \lambda_i f_i\right) P^{\sum_{i=1}^{2} \lambda_i X_i}(dx) = E\left[\varphi\left(0, \sum_{i=1}^{2} \lambda_i X_i; \sum_{i=1}^{2} \lambda_i f_i\right) \right]$$
$$\le E\left[\sum_{i=1}^{2} \lambda_i \left\{ \varphi_i(X_i) + C|X_i|^2 \right\} \right] - CE\left[\left| \sum_{i=1}^{2} \lambda_i X_i \right|^2 \right].$$

In the inequality above, we considered as follows:

$$\varphi(t, x; f) = \log\left(\int_{\mathbb{R}^d} p(t, x; 1, y) \exp(f(y)) dy \right), \; (t, x) \in [0, 1) \times \mathbb{R}^d, \tag{3.46}$$

$$\int_{\mathbb{R}^d} \exp\left(\log p\left(0, \sum_{i=1}^{2} \lambda_i X_i; 1, y\right) + C\left| \sum_{i=1}^{2} \lambda_i X_i \right|^2 + \sum_{i=1}^{2} \lambda_i f_i(y) \right) dy$$
$$\le \int_{\mathbb{R}^d} \exp\left(\sum_{i=1}^{2} \lambda_i \{\log p(0, X_i; 1, y) + C|X_i|^2 + f_i(y)\} \right) dy$$
$$\le \prod_{i=1}^{2} \left(\int_{\mathbb{R}^d} \exp(\log p(0, X_i; 1, y) + C|X_i|^2 + f_i(y)) dy \right)^{\lambda_i}$$

by Hölder's inequality. Taking the supremum in f_i over $C_b^\infty(\mathbb{R}^d)$ on the left-hand side of (3.45), the Duality Theorem for $V_S(P^{X_i}, Q)$ (see Corollary 2.3) completes the proof.

Proof of Corollary 3.3 Let $X, Y \in L^2(\Omega, P; \mathbb{R}^d)$ and $\lambda := \min(1, ||X - Y||_2)$. Then

$$V_S(P^X, Q) - V_S(P^Y, Q) \le \lambda\{2C_2(||X||_2^2 + ||x||_{L^2(Q)}^2) + 2C_1 + C\}, \tag{3.47}$$

where $||X||_2 := \{E[|X|^2]\}^{1/2}$, $X \in L^2(\Omega, P; \mathbb{R}^d)$ (see (3.19) for notation). We prove (3.47) in the case where $\lambda > 0$. Since $Y = (1-\lambda)X + \lambda(\lambda^{-1}(Y - X) + X)$,

$$\begin{aligned}&(1-\lambda)V_S(P^X, Q) + \lambda V_S(P^{\lambda^{-1}(Y-X)+X}, Q)\\ &\le \lambda(1-\lambda)C||\lambda^{-1}(Y-X)||_2^2 + V_S(P^Y, Q)\end{aligned}$$

from Theorem 3.4. From this,

$$\begin{aligned}&V_S(P^X, Q) - V_S(P^Y, Q) \\ &\le \lambda\{V_S(P^X, Q) - V_S(P^{\lambda^{-1}(Y-X)+X}, Q) + C(1-\lambda)||\lambda^{-1}(Y-X)||_2^2\}.\end{aligned} \tag{3.48}$$

Since (A.4)' implies (A.6)–(A.7),

$$\begin{aligned}&V_S(P^X, Q) - V_S(P^{\lambda^{-1}(Y-X)+X}, Q)\\ &\le \mathscr{S}(Q) + C_1 + 2C_2(||X||_2^2 + ||x||_{L^2(Q)}^2) - \mathscr{S}(Q) + C_1\end{aligned}$$

from Proposition 3.3 and (3.19). The following completes the proof of the first part:

$$(1-\lambda)||\lambda^{-1}(Y-X)||_2^2 = \begin{cases}1-\lambda, & \lambda = ||X - Y||_2 < 1,\\ 0 = 1-\lambda, & \lambda = 1.\end{cases}$$

We prove the second part. One can set $C = (2a)^{-1}$. From (3.48), the following holds:

$$\begin{aligned}&V_S(P^X, Q) - V_S(P^Y, Q)\\ &\le \lambda\left\{V_S(P^X, Q) - V_S(P^{\lambda^{-1}(Y-X)+X}, Q) - \frac{1}{2a}||X||_2^2 + \frac{1}{2a}||\lambda^{-1}(Y-X) + X||_2^2\right\}\\ &\quad + \frac{1}{2a}(||X||_2^2 - ||Y||_2^2),\end{aligned}$$

since

$$\lambda(1-\lambda)||\lambda^{-1}(Y-X)||_2^2 = \lambda(-||X||_2^2 + ||\lambda^{-1}(Y-X)+X||_2^2) + ||X||_2^2 - ||Y||_2^2.$$

From Proposition 3.3,

$$\begin{aligned}
&V_S(P^X, Q) - V_S(P^{\lambda^{-1}(Y-X)+X}, Q) - \frac{1}{2a}||X||_2^2 + \frac{1}{2a}||\lambda^{-1}(Y-X)+X||_2^2 \\
&\le \frac{1}{a}\int_{\mathbb{R}^d\times\mathbb{R}^d} \langle x, y\rangle \left\{\mu(P^{\lambda^{-1}(Y-X)+X}, Q)(dxdy) - P^X(dx)Q(dy)\right\} \\
&= \frac{1}{a}\int_{\mathbb{R}^d\times\mathbb{R}^d} \langle x - E[X], y\rangle \mu(P^{\lambda^{-1}(Y-X)+X}, Q)(dxdy) \\
&\le \frac{1}{a\lambda}(||X-Y||_2 + \lambda V(X)^{1/2})||x||_{L^2(Q)},
\end{aligned}$$

which completes the proof.

Proof of Proposition 3.5 From Corollary 3.2, $V_S(P, Q)$ is finite since $\mathcal{S}(Q)$ is finite. There exists $f \in L^1(\mathbb{R}^d, Q)$ such that $\varphi(0, x; f) \in L^1(\mathbb{R}^d, P)$ and that

$$\mu(dxdy; P, Q) = P(dx)p(0, x; 1, y)\exp(f(y) - \varphi(0, x; f))dy \tag{3.49}$$

(see, e.g. [135] and also Corollary 2.3, Theorem 3.3 and Proposition 3.1). In particular,

$$\begin{aligned}
&-V_S(P, Q) \qquad\qquad (3.50)\\
&= -\int_{\mathbb{R}^d} f(y)Q(dy) + \int_{\mathbb{R}^d} \varphi(0, x; f)P(dx) \\
&= \int_{\mathbb{R}^d\times\mathbb{R}^d} (-f(y) + \varphi(0, x; f))P(dx)Q(dy) \\
&= \int_{\mathbb{R}^d\times\mathbb{R}^d} P(dx)Q(dy)\left\{\log\left(\frac{P(dx)Q(dy)}{\mu(dxdy; P, Q)}\right) - \log q(y) + \log p(0, x; 1, y)\right\},
\end{aligned}$$

which completes the proof of (3.40). $P \times Q \mapsto H(P \times Q|\mu(P, Q))$ is weakly lower semicontinuous since $P \times Q \mapsto \mu(P, Q)$ is continuous from Theorem 3.6 in Sect. 3.1.5.2 given later and since $(\mu, \nu) \mapsto H(\mu|\nu)$ is weakly lower semicontinuous (see [52, Lemma 1.4.3]). In particular, (3.41) is weakly lower semicontinuous. The weak lower semicontinuity of (3.41) implies the upper semicontinuity of (3.42) since $L^2(\Omega, P; \mathbb{R}^d) \ni X \mapsto P^X \in \mathcal{P}(\mathbb{R}^d)$ is continuous. (3.42) is also lower semicontinuous by Proposition 2.1 and Lemma 2.2, which completes the proof.

3.1.5 Regularity Result for Schrödinger's Func. Eqn.

In this section, we show that the solution of Schrödinger's Func. Eqn. is measurable in space, kernel, and marginals. As an application, we show that the drift vector of the h-path process with given two endpoint marginals is a measurable function of space, time, and marginal at each time. In particular, we show that the coefficients of mean field PDE systems that the marginals satisfy are measurable functions of space, time, and marginal. This implies a possible development of the SOT in the framework of mean field PDEs and distribution dependent SDEs (see [7] and [27], e.g. [132] and the references therein). Theorem 3.6 in Sect. 3.1.5.2 is used in the proof of Proposition 3.5 in Sect. 3.1.4.

Let S be a σ-compact metric space and $q \in C(S \times S; (0, \infty))$. We recall Schrödinger's Func. Eqn. in a general setting.

Definition 3.3 (Schrödinger's Func. Eqn.) For $\mu_1, \mu_2 \in \mathscr{P}(S)$, find a product measure $\nu_1(dx_1)\nu_2(dx_2)$ of nonnegative σ-finite Borel measures on S for which the following holds:

$$\begin{cases} \mu_1(dx_1) = \nu_1(dx_1) \int_S q(x_1, x_2)\nu_2(dx_2), \\ \\ \mu_2(dx_2) = \nu_2(dx_2) \int_S q(x_1, x_2)\nu_1(dx_1). \end{cases} \tag{3.51}$$

It is known that (3.51) has the unique solution (see [36, 79], and also [13, 63]).

$$u_i(x_i) := \log\left(\int_S q(x_1, x_2)\nu_j(dx_j)\right), \quad i, j = 1, 2, i \neq j. \tag{3.52}$$

Then $\exp(u_1(x))$ and $\exp(u_2(x))$ are positive and

$$\mu_i(dx) = \exp(u_i(x))\nu_i(dx), \quad i = 1, 2. \tag{3.53}$$

Equation (3.51) can be rewritten as follows: for $i, j = 1, 2, i \neq j$,

$$\exp(u_i(x_i)) = \int_S q(x_1, x_2)\exp(-u_j(x_j))\mu_j(dx_j), \quad \mu_i(dx_i)\text{–a.s.}. \tag{3.54}$$

In particular, Schrödinger's Func. Eqn. (3.51) is equivalent to finding a function $u_1(x_1) + u_2(x_2)$ for which (3.54) holds (see Remarks 1.3 and 1.4). Since $\nu_1(dx_1)\nu_2(dx_2)$ is the unique solution of (3.51), it is a functional of μ_1, μ_2, and q. This does not imply the uniqueness of ν_1 and ν_2. Indeed, for $C > 0$,

$$\nu_1 \times \nu_2 = C\nu_1 \times C^{-1}\nu_2.$$

Let $\{K_n\}_{n\geq 1}$ be a nondecreasing sequence of compact subsets of S such that $S = \cup_{n\geq 1} K_n$. $K_1 := S$ in the case where S is compact. We assume that the following holds so that $\nu_i, u_i, i = 1, 2$ are unique:

$$\nu_1(K_{n_0(\mu_1,\mu_2)}) = \nu_2(K_{n_0(\mu_1,\mu_2)}), \tag{3.55}$$

where

$$n_0(\mu_1, \mu_2) := \min\{n \geq 1 | \mu_1(K_n)\mu_2(K_n) > 0\}.$$

It is easy to see that ν_i and u_i are functionals of μ_1, μ_2, and q:

$$\nu_i(dx) = \nu_i(dx; q, \mu_1, \mu_2), \quad u_i(x) = u_i(x; q, \mu_1, \mu_2), \quad i = 1, 2. \tag{3.56}$$

Remark 3.5 If (3.55) holds, then for $i, j = 1, 2, i \neq j$,

$$\nu_i(B) = \frac{\nu_i(B)\nu_j(K_{n_0(\mu_1,\mu_2)})}{\{\nu_1(K_{n_0(\mu_1,\mu_2)})\nu_2(K_{n_0(\mu_1,\mu_2)})\}^{1/2}}. \tag{3.57}$$

This implies that the measurability of $(q, \mu_1, \mu_2) \mapsto \nu_i(dx; q, \mu_1, \mu_2)$ can be reduced to that of $(q, \mu_1, \mu_2) \mapsto \nu_1(dx; q, \mu_1, \mu_2)\nu_2(dx; q, \mu_1, \mu_2)$.

Let $\mathcal{M}(S)$ denote the space of all Radon measures on S. We denote by a Radon measure a locally finite and inner regular Borel measure. It is known that a locally finite and σ-finite Borel measure on a σ-compact metric space is a Radon measure in our sense (see, e.g. [68, p. 901, Prop. 32.3.4]).

We first consider the case where $\mathcal{P}(S)$ is endowed with a strong topology. In Theorem 3.5, we show that if S is compact, then the following is strongly continuous:

$$\nu_i(dx; \cdot, \cdot, \cdot) : C(S \times S) \times \mathcal{P}(S) \times \mathcal{P}(S) \longrightarrow \mathcal{M}(S),$$

$$u_i : C(S \times S) \times \mathcal{P}(S) \times \mathcal{P}(S) \longrightarrow C(S),$$

and $u_i \in C(S \times C(S \times S) \times \mathcal{P}(S) \times \mathcal{P}(S))$. In Corollary 3.4, we also show that if S is σ-compact, then the following is weakly Borel measurable and Borel measurable, respectively:

$$\nu_i(dx; \cdot, \cdot, \cdot) : C(S \times S) \times \mathcal{P}(S) \times \mathcal{P}(S) \longrightarrow \mathcal{M}(S),$$

$$u_i : S \times C(S \times S) \times \mathcal{P}(S) \times \mathcal{P}(S) \longrightarrow \mathbb{R} \cup \{\infty\}.$$

As an application of this measurability result, we show that the coefficients of the mean field PDE system which the marginal distributions of the h-path process with given two endpoint marginals satisfy are measurable functions of space, time, and marginal. More precisely, in the setting of Theorem 3.3 in Sect. 3.1.1, the following holds (see (3.52) and (3.56) for notation): for $x \in \mathbb{R}^d$,

$$h(t,x) = \begin{cases} \exp\{u_1(x; p(t,\cdot;1,\cdot), P^{X(t)}, P_1)\}, & t \in [0,1), \\ \dfrac{\nu_2(dx; p(0,\cdot;t,\cdot), P_0, P^{X(t)})}{dx} & \\ = \exp\{-u_2(x; p(0,\cdot;t,\cdot), P_0, P^{X(t)})\} \dfrac{P^{X(t)}(dx)}{dx}, & t \in (0,1]. \end{cases} \tag{3.58}$$

As an application of Corollary 3.4, we show that

$$U(t,x,P) := u_1(x; p(t,\cdot;1,\cdot), P, P_1), \quad (t,x,P) \in [0,1) \times \mathbb{R}^d \times \mathscr{P}(\mathbb{R}^d) \tag{3.59}$$

is a Borel measurable function from $[0,1) \times \mathbb{R}^d \times \mathscr{P}(\mathbb{R}^d)$ to $\mathbb{R}$ (see Corollary 3.5). Theorems 3.2 and 3.3, and (3.58)–(3.59) imply that if $P_1(dy) \ll dy$, then $P^{X(t)}(dx) \ll dx, t \in (0,1]$ and $P^{X(t)}(dx)$ satisfies the following mean field PDE system (see [10, 11, 27, 93], and the references therein for the mean field games and the master equations). For any $f \in C_b^2(\mathbb{R}^d)$ and $t \in [0,1]$,

$$\begin{aligned} &\int_{\mathbb{R}^d} f(x)P^{X(t)}(dx) - \int_{\mathbb{R}^d} f(x)P_0(dx) \\ &= \int_0^t ds \int_{\mathbb{R}^d} (A_s f(x) + \langle a(s,x)D_x U(s,x,P^{X(s)}), Df(x)\rangle) P^{X(s)}(dx), \end{aligned} \tag{3.60}$$

$$\begin{aligned} 0 = {} & \partial_t U(t,x,P^{X(t)}) + A_t U(t,x,P^{X(t)}) \\ & + \frac{1}{2}\langle a(t,x)D_x U(t,x,P^{X(t)}), D_x U(t,x,P^{X(t)})\rangle, \quad (t,x) \in (0,1) \times \mathbb{R}^d \end{aligned} \tag{3.61}$$

$$U(1,x,P_1) = \log\left(\frac{\nu_2(dx; p(0,\cdot;1,\cdot), P_0, P_1)}{dx}\right), \quad P_1(dx)\text{–a.e.}$$

(see (3.4) for notation). Here we consider $U(t,x,P^{X(t)})$ a function of (t,x).

We also consider the case where S is also complete and $\mathscr{P}(S)$ is endowed with weak topology. We refer readers to our recent papers [116] and [117] for the proof of the case where $\mathscr{P}(S)$ is endowed with a strong topology and for an application of our result in the case where $\mathscr{P}(S)$ is endowed with weak topology, respectively.

3.1.5.1 Regularity of Schrödinger's Func. Eqn. in the Strong Topology

In this section, we only state our result [116] where $\mathscr{P}(S)$ is endowed with a strong topology. We refer readers to [116] for the proof.

We first describe assumptions:

(A.8). S is a compact metric space.
(A.9). $q \in C(S \times S; (0, \infty))$.
(A.8)'. S is a σ-compact metric space.

For a metric space X and $\mu \in \mathscr{M}(X)$,

$$||\mu|| := \sup\left\{\int_X \phi(x)\mu(dx)\middle|\phi \in C(X), ||\phi||_\infty \leq 1\right\} \in [0, \infty]. \tag{3.62}$$

In the case where S is compact, we have the continuity results on ν_i, u_i in (3.56) (recall (3.55)).

Theorem 3.5 (See [116]) *Suppose that (A.8) and (A.9) hold. Suppose also that* $\mu_{i,n}, \mu_i \in \mathscr{P}(S)$, $q_n \in C(S \times S; (0, \infty))$, $i = 1, 2$, $n \geq 1$ *and*

$$\lim_{n\to\infty} (||\mu_{1,n} \times \mu_{2,n} - \mu_1 \times \mu_2|| + ||q_n - q||_\infty) = 0. \tag{3.63}$$

Then

$$\lim_{n\to\infty} ||\nu_1(\cdot; q_n, \mu_{1,n}, \mu_{2,n}) \times \nu_2(\cdot; q_n, \mu_{1,n}, \mu_{2,n}) \tag{3.64}$$
$$- \nu_1(\cdot; q, \mu_1, \mu_2) \times \nu_2(\cdot; q, \mu_1, \mu_2)|| = 0,$$

$$\lim_{n\to\infty} \sum_{i=1}^{2} ||u_i(\cdot; q_n, \mu_{1,n}, \mu_{2,n}) - u_i(\cdot; q, \mu_1, \mu_2)||_\infty = 0. \tag{3.65}$$

Besides, for $i = 1, 2$, *and* $\{x_n\}_{n\geq 1} \subset S$ *which converges, as* $n \to \infty$, *to* $x \in S$,

$$\lim_{n\to\infty} u_i(x_n; q_n, \mu_{1,n}, \mu_{2,n}) = u_i(x; q, \mu_1, \mu_2). \tag{3.66}$$

In the case where S is σ-compact, we only have the Borel measurability results on ν_i, u_i.

Corollary 3.4 (See [116]) *Suppose that (A.8)' and (A.9) hold. Then the following is Borel measurable: for* $i = 1, 2$,

$$\int_S f(x)\nu_i(dx; \cdot, \cdot, \cdot) : C(S \times S) \times \mathscr{P}(S) \times \mathscr{P}(S) \longrightarrow \mathbb{R}, \quad f \in C_0(S),$$

$$u_i : S \times C(S \times S) \times \mathscr{P}(S) \times \mathscr{P}(S) \longrightarrow \mathbb{R} \cup \{\infty\}.$$

As an application of Corollary 3.4, the following holds.

Corollary 3.5 (See [116]) *Suppose that (A.6) and (A.7) hold. Then $U(t, x, P)$ in (3.59) is a Borel measurable function from $[0, 1) \times \mathbb{R}^d \times \mathscr{P}(\mathbb{R}^d)$ to $\mathbb{R}$. In particular, (3.60)–(3.61) hold.*

3.1.5.2 Regularity of Schrödinger's Func. Eqn. in the Weak Topology

In this section, we state and prove our measurability result in [117] where $\mathscr{P}(S)$ is endowed with weak topology.

We first describe another assumption:

(A.8)''. S is a complete σ-compact metric space.

We recall that $C(S \times S)$ is endowed with the topology induced by the uniform convergence on every compact subset of S.

Under (A.8)'', let $\{\varphi_m\}_{m \geq 1}$ be a nondecreasing sequence of functions in $C_0(S; [0, 1])$ such that the following holds:

$$\varphi_m(x) = 1, \quad x \in K_m, m \geq 1$$

(see (3.55)). If $S = \mathbb{R}^d$, then $K_m := B_m$ and we assume that $\varphi_m \in C_0(B_{m+1}; [0, 1])$. For $i \neq j, i, j = 1, 2, m \geq 1$, and $x_i \in S$,

$$u_{i|m}(x_i; q, \mu_1, \mu_2) := \log \left(\int_S q(x_1, x_2)\varphi_m(x_j)\nu_j(dx_j; q, \mu_1, \mu_2) \right), \tag{3.67}$$

provided the right-hand side is well defined (see (3.56) and also (3.52)).

$$\mu(dxdy; q, \mu_1, \mu_2) := \nu_1(dx; q, \mu_1, \mu_2)q(x, y)\nu_2(dx; q, \mu_1, \mu_2). \tag{3.68}$$

The following is the continuity result on $\nu_1 \times \nu_2$, μ, and $u_{i|m}$.

Theorem 3.6 (See [117]) *Suppose that (A.8)'' and (A.9) hold and that $q_n \in C(S \times S; (0, \infty))$, $\mu_i, \mu_{i,n} \in \mathscr{P}(S)$, $n \geq 1$, $i = 1, 2$, and*

$$\lim_{n \to \infty} q_n = q, \quad \textit{locally uniformly}, \tag{3.69}$$

$$\lim_{n \to \infty} \mu_{1,n} \times \mu_{2,n} = \mu_1 \times \mu_2, \quad \textit{weakly}. \tag{3.70}$$

Then for any $f \in C_0(S \times S)$,

$$\lim_{n \to \infty} \int_{S \times S} f(x, y)\nu_1(dx; q_n, \mu_{1,n}, \mu_{2,n})\nu_2(dy; q_n, \mu_{1,n}, \mu_{2,n}) \tag{3.71}$$

$$= \int_{S \times S} f(x, y)\nu_1(dx; q, \mu_1, \mu_2)\nu_2(dy; q, \mu_1, \mu_2).$$

In particular,

$$\lim_{n\to\infty} \mu(dxdy; q_n, \mu_{1,n}, \mu_{2,n}) = \mu(dxdy; q, \mu_1, \mu_2), \quad \textit{weakly.} \tag{3.72}$$

For any $\{x_{i,n}\}_{n\geq 1} \subset S$ *which converges, as* $n \to \infty$, *to* $x_i \in S$, $i = 1, 2$ *and for sufficiently large* $m \geq 1$,

$$\lim_{n\to\infty} \sum_{i=1}^{2} u_{i|m}(x_{i,n}; q_n, \mu_{1,n}, \mu_{2,n}) = \sum_{i=1}^{2} u_{i|m}(x_i; q, \mu_1, \mu_2). \tag{3.73}$$

$n_0(\mu_1, \mu_2) \leq n$ if and only if $\mu_1(K_n)\mu_2(K_n) > 0$ (see (3.55) for notation). $(\mu_1, \mu_2) \mapsto \mu_1(K_n)\mu_2(K_n)$ is upper semicontinuous in the weak topology since a compact set is closed. In particular, $(\mu_1, \mu_2) \mapsto n_0(\mu_1, \mu_2)$ is measurable and Theorem 3.6 implies the following (see [116, proof of Corollary 2.1]).

Corollary 3.6 (See [117]) *Suppose that (A.8)" and (A.9) hold. Then the following is Borel measurable: for* $i = 1, 2$,

$$\int_S f(x)\nu_i(dx; \cdot, \cdot, \cdot) : C(S \times S) \times \mathscr{P}(S) \times \mathscr{P}(S) \longrightarrow \mathbb{R}, \quad f \in C_0(S),$$

$$u_i : S \times C(S \times S) \times \mathscr{P}(S) \times \mathscr{P}(S) \longrightarrow \mathbb{R} \cup \{\infty\}.$$

If S is compact, then $\nu_1(S) = \nu_2(S)$ (see (3.55)). This implies the following from Theorem 3.6.

Corollary 3.7 (See [117]) *Suppose that (A.8) and the assumptions of Theorem 3.6 except (A.8)" hold. Then the following holds: for* $i = 1, 2$,

$$\lim_{n\to\infty} \int_S f(x)\nu_i(dx; q_n, \mu_{1,n}, \mu_{2,n}) = \int_S f(x)\nu_i(dx; q, \mu_1, \mu_2), \quad f \in C(S),$$

and for any $\{x_n\}_{n\geq 1} \subset S$ *which converges, as* $n \to \infty$, *to* $x \in S$,

$$\lim_{n\to\infty} u_i(x_n; q_n, \mu_{1,n}, \mu_{2,n}) = u_i(x; q, \mu_1, \mu_2).$$

A uniformly bounded sequence of convex functions on a convex neighborhood N_A of a convex subset A of $\mathbb{R}^d$ is compact in $C(A)$, provided $dist(A, N_A^c)$ is positive (see, e.g. [8, section 3.3]). We describe an additional assumption and state a stronger result than above, provided $S \subset \mathbb{R}^d$.

(A.9.*r*)'. There exists $C_r > 0$ for which $x \mapsto C_r|x|^2 + \log q(x, y)$ and $y \mapsto C_r|y|^2 + \log q(x, y)$ are convex on B_r for any $y \in B_r$ and any $x \in B_r$, respectively.

Remark 3.6 If $\log q(x, y)$ has bounded second-order partial derivatives on B_r, then (A.9.r)' holds (see Remark 3.2, (i) in Sect. 3.1.4).

$$||f||_{\infty,r} := \sup_{x \in B_r} |f(x)|, \quad f \in C(B_r). \tag{3.74}$$

The following is a stronger convergence result than Corollary 3.7 (see Lemma 3.1 below).

Corollary 3.8 (See [117]) *Let $r > 0$. Suppose that (A.9.r)' and the assumptions of Corollary 3.7 with $S = B_r$ hold. Then for any $r' < r$,*

$$\lim_{n \to \infty} \sum_{i=1}^{2} ||u_i(\cdot; q_n, \mu_{1,n}, \mu_{2,n}) - u_i(\cdot; q, \mu_1, \mu_2)||_{\infty,r'} = 0. \tag{3.75}$$

We state a lemma and prove Theorem 3.6. The following can be proved easily.

Lemma 3.1 *Suppose that (A.8)" and (A.9) hold. Then for any $\mu_1, \mu_2 \in \mathscr{P}(S)$, μ defined by (3.68),*

$$\begin{aligned} \min_{x,y \in K_m} q(x, y)^{-1} \mu(K_m \times K_m) &\le \int_S \varphi_m(x) \nu_1(dx) \int_S \varphi_m(y) \nu_2(dy) \\ &\le \max_{x,y \in supp(\varphi_m)} q(x, y)^{-1}. \end{aligned} \tag{3.76}$$

Proof of Theorem 3.6 We first prove (3.71). For the sake of simplicity,

$$\begin{aligned} \nu_{i,n}(dx) &:= \nu_i(dx; q_n, \mu_{1,n}, \mu_{2,n}), \\ \mu_n(dxdy) &:= \nu_{1,n}(dx) q_n(x, y) \nu_{2,n}(dy). \end{aligned} \tag{3.77}$$

Since $\{\mu_{1,n}(dx) = \mu_n(dx \times S), \mu_{2,n}(dy) = \mu_n(S \times dy)\}_{n \ge 1}$ is convergent, $\{\mu_n\}_{n \ge 1}$ is tight. Indeed, for any Borel sets $A, B \in \mathbb{R}^d$,

$$\mu_n((A \times B)^c) \le \mu_n(A^c \times S) + \mu_n(S \times B^c) = \mu_{1,n}(A^c) + \mu_{2,n}(B^c),$$

and a convergent sequence of probability measures on a complete separable metric space is tight by Prohorov's Theorem (see, e.g. [14]). Here notice that a σ-compact metric space is separable. By Prohorov's Theorem, take a weakly convergent subsequence $\{\mu_{n_k}\}_{k \ge 1}$ and denote the limit by μ. Then it is easy to see that the following holds:

$$\mu(dx \times S) = \mu_1(dx), \quad \mu(S \times dy) = \mu_2(dy). \tag{3.78}$$

From (A.9) and (3.69)–(3.70), the following holds: for any $f \in C_0(S \times S)$,

$$\lim_{k\to\infty} \int_{S\times S} f(x, y)\nu_{1,n_k}(dx)\nu_{2,n_k}(dy) = \int_{S\times S} f(x, y)q(x, y)^{-1}\mu(dxdy). \tag{3.79}$$

Indeed,

$$\nu_{1,n}(dx)\nu_{2,n}(dy) = \left(\frac{1}{q_n(x, y)} - \frac{1}{q(x, y)}\right)\mu_n(dxdy) + \frac{1}{q(x, y)}\mu_n(dxdy).$$

The rest of the proof of (3.71) is divided into (3.80)–(3.81) below, which will be proved later.

There exists a subsequence $\{\overline{n}_k\} \subset \{n_k\}$ and finite measures $\overline{\nu}_{1,m}, \overline{\nu}_{2,m} \in \mathcal{M}(supp(\varphi_m))$ such that for sufficiently large $m \geq 1$ and any $f \in C_0(S \times S)$,

$$\lim_{k\to\infty} \int_{S\times S} f(x, y)\varphi_m(x)\varphi_m(y)\nu_{1,\overline{n}_k}(dx)\nu_{2,\overline{n}_k}(dy) \tag{3.80}$$
$$= \int_{S\times S} f(x, y)\overline{\nu}_{1,m}(dx)\overline{\nu}_{2,m}(dy).$$

From (3.80), for sufficiently large $m \geq 1$ and any Borel sets $A_1, A_2 \subset S$,

$$\int_{A_1\times A_2} q(x, y)^{-1}\mu(dxdy) \tag{3.81}$$
$$= \frac{\int_{A_1\times K_m} q(x, y)^{-1}\mu(dxdy) \int_{K_m\times A_2} q(x, y)^{-1}\mu(dxdy)}{\overline{\nu}_{1,m}(K_m)\overline{\nu}_{2,m}(K_m)}.$$

Equations (3.78) and (3.81) imply that $q(x, y)^{-1}\mu(dxdy)$ is a product measure which satisfies (3.51). Equation (3.79) and the uniqueness of the solution to (3.51) implies that (3.71) is true.

We prove (3.80)–(3.81) to complete the proof of (3.71). Equation (3.80) can be proved by the diagonal method, since $\{\mu_n\}_{n\geq 1}$ is tight and since for sufficiently large $m \geq 1$,

$$\varphi_m(x_1)\varphi_m(x_2)\nu_{1,n_k}(dx_1)\nu_{2,n_k}(dx_2) \tag{3.82}$$
$$= \int_S \varphi_m(x)\nu_{1,n_k}(dx) \int_S \varphi_m(y)\nu_{2,n_k}(dy) \frac{\varphi_m(x_1)\nu_{1,n_k}(dx_1)}{\int_S \varphi_m(x)\nu_{1,n_k}(dx)} \frac{\varphi_m(x_2)\nu_{2,n_k}(dx_2)}{\int_S \varphi_m(x)\nu_{2,n_k}(dx)}$$

has a convergent subsequence from (3.76) in Lemma 3.1 by Prohorov's Theorem and since any weak limit is a product measure. We prove (3.81). From (3.79) and (3.80), for sufficiently large $\tilde{m} \geq 1$,

$$\begin{aligned} &\int_{(A_1\times A_2)\cap(K_{\tilde{m}}\times K_{\tilde{m}})} q(x,y)^{-1}\mu(dxdy) \\ &= \int_{(A_1\times A_2)\cap(K_{\tilde{m}}\times K_{\tilde{m}})} \varphi_{\tilde{m}}(x)\varphi_{\tilde{m}}(y)q(x,y)^{-1}\mu(dxdy) \\ &= \int_{(A_1\times A_2)\cap(K_{\tilde{m}}\times K_{\tilde{m}})} \overline{\nu}_{1,\tilde{m}}(dx)\overline{\nu}_{2,\tilde{m}}(dy) \\ &= \overline{\nu}_{1,\tilde{m}}(A_1\cap K_{\tilde{m}})\overline{\nu}_{2,\tilde{m}}(A_2\cap K_{\tilde{m}}). \end{aligned} \tag{3.83}$$

From (3.83), for $\tilde{m} \geq m$, setting $A_i = K_m$,

$$\begin{aligned} \overline{\nu}_{1,\tilde{m}}(A_1\cap K_{\tilde{m}}) &= \frac{\int_{(A_1\times K_m)\cap(K_{\tilde{m}}\times K_{\tilde{m}})} q(x,y)^{-1}\mu(dxdy)}{\overline{\nu}_{2,\tilde{m}}(K_m)}, \\ \overline{\nu}_{2,\tilde{m}}(A_2\cap K_{\tilde{m}}) &= \frac{\int_{(K_m\times A_2)\cap(K_{\tilde{m}}\times K_{\tilde{m}})} q(x,y)^{-1}\mu(dxdy)}{\overline{\nu}_{1,\tilde{m}}(K_m)}, \\ \overline{\nu}_{1,\tilde{m}}(K_m)\overline{\nu}_{2,\tilde{m}}(K_m) &= \overline{\nu}_{1,m}(K_m)\overline{\nu}_{2,m}(K_m) = \int_{K_m\times K_m} q(x,y)^{-1}\mu(dxdy). \end{aligned} \tag{3.84}$$

Substitute (3.84) for (3.83) and let $\tilde{m} \to \infty$. Then we obtain (3.81). Equation (3.73) can be shown from (3.71) by the following: from (3.67),

$$\begin{aligned} &\exp\left(\sum_{i=1}^{2} u_{i|m}(x_{i,n}; q_n, \mu_{1,n}, \mu_{2,n})\right) \\ &= \int_{S\times S} q_n(x_{1,n}, y)q_n(x, x_{2,n})\varphi_m(x)\varphi_m(y)\nu_{1,n}(dx)\nu_{2,n}(dy), \end{aligned} \tag{3.85}$$

provided the right-hand side is positive.

3.2 Marginal Problem for Stochastic Processes

The problem of the construction of a stochastic process with given marginal distributions is an important part of the so-called marginal problem (see [82, 86, 113, 130, 142], and the references therein). This is not the only problem in the theory of probability. For instance, consider the construction of a graph of a function with a given Gauss curvature. For a smooth function $u : \mathbb{R}^d \longrightarrow \mathbb{R}$, consider the interior of the epigraph epi (u) of u as the inside of the surface $y = u(x)$:

$$\text{epi }(u) := \{(x, y)|x \in \mathbb{R}^d, y \geq u(x)\}.$$

Then the Gauss curvature $\mathscr{K}(x, u(x))$ at $(x, u(x))$ is given by the following:

$$\mathscr{K}(x, u(x)) := (1 + |Du(x)|^2)^{-\frac{d+2}{2}} \det(D^2u(x)), \quad x \in \mathbb{R}^d. \tag{3.86}$$

As a problem of the PDE, the problem is reduced to the following Monge–Ampére equation: for a probability density function $p : \mathbb{R}^d \longrightarrow [0, \infty)$,

$$p(x) = \tilde{p}_C(Du(x))\det(D^2u(x)), \quad x \in \mathbb{R}^d, \tag{3.87}$$

where

$$\tilde{p}_C(x) := \frac{(1 + |x|^2)^{-\frac{1}{2}} p_C(x)}{\int_{\mathbb{R}^d} (1 + |y|^2)^{-\frac{1}{2}} p_C(y)dy}, \quad x \in \mathbb{R}^d.$$

Here p_C denotes the probability density function of a Cauchy distribution:

$$p_C(x) := \Gamma\left(\frac{d+1}{2}\right)\pi^{-\frac{d+1}{2}}(1 + |x|^2)^{-\frac{d+1}{2}}, \quad x \in \mathbb{R}^d.$$

This research was originated by A. D. Alexandrov, A. V. Pogorerov, I. J. Bakleman et al., and is still actively continued by L. A. Caffarelli et al. (see, e.g. [8, 72, 108], and the references therein). It can be considered, from the probability theory point of view, as follows. Construct a random variable X defined on the probability space $(\mathbb{R}^d, \mathbf{B}(\mathbb{R}^d), \tilde{p}_C(x)dx)$ such that there exists a function u for which the following holds:

$$(\tilde{p}_C(x)dx)^X = p(x)dx, \quad X(x) = (Du)^{-1}(x), \quad p_C(x)dx\text{–a.e..}$$

In the case where $P_t \in \mathscr{P}(\mathbb{R}^d)$ is given at each time $t \in [0, 1]$, R. M. Blumenthal and H. H. Corson constructed a Borel probability measure on $C([0, 1]; \mathbb{R}^d)$ with a marginal distribution P_t at each time $t \in [0, 1]$ (see [15]). But they did not study stochastic dynamics. In this section, given marginal distributions P_t at $t = 0, 1$ or at each time $t \in [0, 1]$, we discuss the variational approach to the construction of a Borel probability measure on $C([0, 1]; \mathbb{R}^d)$ as the probability distribution of a solution to a SDE or an ODE. We also consider the Markov property of solutions to the SOTs with a nonconvex cost in the one-dimensional case as an application of the superposition principle and the one-dimensional OT with a concave cost.

3.2.1 Marginal Problem for SDEs

In this section, we discuss a variational approach to the construction of a solution to the SDE with given marginal distributions.

The square of the absolute value of a solution to Schrödinger's equation is the probability density of a quantum particle by Born's probabilistic interpretation of a solution to Schrödinger's equation, provided it is normalized. E. Nelson proposed the problem of the construction of a Markov diffusion process of which the marginal distribution at each time is this probability density (see [125, 126]). As a mathematical problem, it is the problem of the construction of a Markov diffusion process from the Fokker–Planck equation.

We write $(a, b) \in \mathbf{A}_0(\{P_t\}_{0\le t\le 1})$ if $a, b \in L^1_{loc}([0,1]\times\mathbb{R}^d, dtP_t(dx))$ and if (2.15) holds for all $f \in C_0^{1,2}([0,1]\times\mathbb{R}^d)$. In the case where a is fixed, we write $b \in \mathbf{A}_0(\{P_t\}_{0\le t\le 1})$ for simplicity. We state a generalized version of Nelson's problem and call it Nelson's problem from a probabilistic point of view (see [106, 107] for the related topics).

Definition 3.4 (Nelson's Problem) For $\{P_t\}_{0\le t\le 1} \subset \mathcal{P}(\mathbb{R}^d)$ such that $\mathbf{A}_0(\{P_t\}_{0\le t\le 1})$ is not empty and for $(a, b) \in \mathbf{A}_0(\{P_t\}_{0\le t\le 1})$, construct a $d\times d$ matrix-valued function $\sigma(t,x)$ on $[0,1]\times\mathbb{R}^d$ and a semimartingale $\{X(t)\}_{0\le t\le 1}$ such that the following holds: for $(t,x) \in [0,1]\times\mathbb{R}^d$,

$$a(t,x) = \sigma(t,x)\sigma(t,x)^t, \quad dtP_t(dx)\text{–a.e.}, \tag{3.88}$$

$$X(t) = X(0) + \int_0^t b(s, X(s))ds + \int_0^t \sigma(s, X(s))dW_X(s), \tag{3.89}$$

$$P^{X(t)} = P_t \tag{3.90}$$

(see Remark 2.3 for notation in (3.89)).

The first result was given by E. Carlen in the case where a is an identity matrix (see [28, 29] and also [31, 123, 159] for different approaches). We generalized it to the case with a variable diffusion matrix (see [103]). P. Cattiaux and C. Léonard extensively generalized it to the case which also includes the construction of a jump-type Markov process (see [32–35]). In these papers they assumed the following:

Definition 3.5 (Finite Energy Condition (FEC)) There exists $b \in \mathbf{A}(\{P_t\}_{0\le t\le 1})$ such that the following holds:

$$\int_0^1 dt \int_{\mathbb{R}^d} \langle a(t,x)^{-1}b(t,x), b(t,x)\rangle P_t(dx) < \infty. \tag{3.91}$$

Remark 3.7 For $\{P_t\}_{0\le t\le 1}$ in (2.15) and for a fixed matrix a, $b \in \mathbf{A}(\{P_t\}_{0\le t\le 1})$ is not necessarily unique (see [103] or [32–35]).

By the continuum limit of $V(\cdot,\cdot)$, we also considered Nelson's problem in a more general setting as follows (see [111]).

Definition 3.6 (Generalized Finite Energy Condition (GFEC)) There exists $\gamma > 1$ and $b \in \mathbf{A}(\{P_t\}_{0\le t\le 1})$ such that

$$\int_0^1 dt \int_{\mathbb{R}^d} \langle a(t,x)^{-1} b(t,x), b(t,x)\rangle^{\frac{\gamma}{2}} P_t(dx) < \infty. \tag{3.92}$$

As an application of the Duality Theorem for $\mathbf{V}$ in Sect. 2.2.2, we gave an approach to Nelson's problem under the GFEC via the SOT (see [112]). We showed the Duality Theorems in Corollary 2.2 and Theorem 2.6, (i) in Sect. 2.2.1 first and then showed Proposition 2.1 in Sect. 2.2 (see [112] and also [103, 111]). Besides, the Duality Theorem for $\mathbf{V}$ implies the functional equation for Nelson's problem which is an analog of Schrödinger's Func. Eqn. for Schrödinger's problem (see Definition 3.8 given later).

D. Trevisan's superposition principle almost completely solved Nelson's problem (3.88)–(3.90) (see Theorem 2.3 in Sect. 2.2 and also [19, 133] for the recent development). We also improved our previous results on the SOTs using that. Besides, it also generalized our approaches to Nelson's problem and to an h-path process via the SOTs though $\mathbf{V}(\{P_t\}_{0\le t\le 1})$ gives an optimal drift vector among $\mathbf{A}(\{P_t\}_{0\le t\le 1})$ for $\{P_t\}_{0\le t\le 1} \subset \mathcal{P}(\mathbb{R}^d)$ and for a fixed matrix a (see Sect. 2.2).

In his problem, E. Nelson considered the case where $a = Identity$ and $b = D_x\psi(t,x)$ for some function ψ. It turned out that it is the minimizer of $\mathbf{V}$ in the case where (2.47) with $a = Identity$, $\xi = 0$, and $c = 0$ and the FEC hold (see [103, Proposition 3.1], and also Theorem 2.6, (ii) in Sect. 2.2.1). Indeed, if $(a, D_x\psi_i) \in \mathbf{A}(\{P_t\}_{0\le t\le 1})$, $i = 1, 2$, then $D_x\psi_1 = D_x\psi_2$, $dtP_t(dx)$–a.e. In this sense, Nelson's original problem can be considered the studies of the superposition principle and of the minimizer of $\mathbf{V}$. More precisely, if the superposition principle holds, then the set over which the infimum is taken in $\mathbf{V}$ is not empty and one can consider a minimizer of $\mathbf{V}$, provided it is finite.

Theorems 3.1–3.3 imply an alternative description of Schrödinger's Func. Eqn. as follows (see Remarks 1.3 and 1.4).

Definition 3.7 (Schrödinger's Func. Eqn.) Suppose that (A.6) and (A.7) hold. For given $P_0 \in \mathcal{P}(\mathbb{R}^d)$ and $P_1 \in \mathcal{P}_{ac}(\mathbb{R}^d)$, find $f \in L^1(\mathbb{R}^d, P_1)$ for which the following holds:

$$P_1(dy) = \left\{ \int_{\mathbb{R}^d} P_0(dx) \frac{p(0,x;1,y)}{E_{0,x}[\exp(f(\mathbf{X}(1)))]} \right\} \exp(f(y))dy. \tag{3.93}$$

The following gives a variational meaning to Schrödinger's Func. Eqn. from the SOT point of view.

Proposition 3.6 (See [116]) *Suppose that (A.6) and (A.7) hold. Then for any $P_0, P_1 \in \mathscr{P}(\mathbb{R}^d)$ for which $V(P_0, P_1)$ is finite, Schrödinger's Func. Eqn. (3.93) is equivalent to the following: find a function $f \in L^1(\mathbb{R}^d, P_1)$ such that*

$$P_1(dy) = \frac{\delta V^*_{P_0}(f)}{\delta f}(dy). \tag{3.94}$$

Here $\frac{\delta V^*_{P_0}(f)}{\delta f}$ denotes the Gâteaux derivative of $V^*_{P_0}(f)$ (see (2.102) for notation).

Proof of Theorem 3.6

$$\frac{\delta V^*_{P_0}(f)}{\delta f}(dy) = \left(\int_{\mathbb{R}^d} P_0(dx) p(0, x; 1, y) \exp(f(y) - \varphi(0, x; f)) \right) dy \tag{3.95}$$

(see (3.46) for notation). Indeed, for any $\psi \in C_b(\mathbb{R}^d)$ and $\varepsilon \in \mathbb{R}$, consider Schrödinger's Func. Eqn. (3.2) with P_1 replaced by $\mu^{\varepsilon\psi}$, where for $\phi \in C_b(\mathbb{R}^d)$,

$$\mu^{\phi}(dy) := \exp(f(y) + \phi(y)) dy \int_{\mathbb{R}^d} P_0(dx) p(0, x; 1, y) \exp(-\varphi(0, x; f + \phi)).$$

Then, from Lemma 2.5 in Sect. 2.2.2,

$$V^*_{P_0}(f + \varepsilon\psi) = \int_{\mathbb{R}^d} \log \left(\int_{\mathbb{R}^d} p(0, x; 1, y) \exp(f(y) + \varepsilon\psi(y)) dy \right) P_0(dx).$$

This implies (3.95). From (3.46),

$$P_0(dx) = P_0(dx) \left(\int_{\mathbb{R}^d} p(0, x; 1, y) \exp(f(y) - \varphi(0, x; f)) dy \right). \tag{3.96}$$

Equations (3.95) and (3.96) complete the proof.

Suppose that (A.6)–(A.7) hold and that (2.47) with $c = 0$ holds. If (2.57) has a maximizer f, then for the minimizer X of $\mathbf{V}(\{P_t\}_{0 \le t \le 1})$,

$$P(X \in B) = E\left[\exp\left(\int_0^1 f(s, \mathbf{X}(s)) ds - \phi(0, \mathbf{X}(0); f) \right); \mathbf{X} \in B \right], \tag{3.97}$$

$B \in \mathbf{B}(C([0, 1]))$. Equation (3.97) implies the following which we do not know how to solve. This is our future problem.

Definition 3.8 (Functional Equation for Nelson's Problem (See [112])) For a given $\{P_t\}_{0\le t\le 1} \subset \mathscr{P}(\mathbb{R}^d)$, find $f \in L^1([0,1]\times\mathbb{R}^d, dtP_t(dx))$ for which the following holds: for $t \in (0,1]$,

$$\frac{P_t(dy)}{dy} = \int_{\mathbb{R}^d} P_0(dx) \frac{p(0,x;t,y)E_{0,x}[\exp(\int_0^1 f(s,\mathbf{X}(s))ds)|(t,\mathbf{X}(t)=y)]}{E_{0,x}[\exp(\int_0^1 f(s,\mathbf{X}(s))ds)]}. \tag{3.98}$$

3.2.2 Marginal Problems for ODEs

In this section, we consider the case where $a = 0$ in (2.15). We only state known results and refer readers to published papers for the proofs.

Stochastic processes under consideration are absolutely continuous in time as in the case of the OT. Our PDE is Liouville's equation, i.e., instead of (2.15), we consider the following: for any $f \in C_b^1([0,1]\times\mathbb{R}^d)$ and $t \in [0,1]$,

$$\begin{aligned}&\int_{\mathbb{R}^d} f(t,x)P_t(dx) - \int_{\mathbb{R}^d} f(0,x)P_0(dx) \\ &= \int_0^t ds \int_{\mathbb{R}^d} \Big(\partial_s f(s,x) + \langle b(s,x), D_x f(s,x)\rangle\Big) P_s(dx).\end{aligned} \tag{3.99}$$

In this section, for the sake of simplicity, we assume the following.

(A.10). $L = L(u)$, $L : \mathbb{R}^d \longrightarrow [0,\infty)$ is convex, and $\liminf_{|u|\to\infty} L(u)/|u|^2 > 0$.

In the case where $\sigma = \sqrt{\varepsilon}\times \mathrm{Id}$, for $\mathbf{P} := \{P_t\}_{0\le t\le 1} \subset \mathscr{P}(\mathbb{R}^d)$, we write $\mathbf{V}_\varepsilon(\mathbf{P}) := \mathbf{V}(\mathbf{P})$ for $\varepsilon > 0$ and $\mathbf{v}_\varepsilon(\mathbf{P}) := \mathbf{v}(\mathbf{P})$ for $\varepsilon \ge 0$ (see Definitions 2.2–2.3 for notation), and

$$\begin{aligned}\mathbf{V}_0(\mathbf{P}) := \inf\Big\{E\Big[\int_0^1 L(\dot{X}(t))dt\Big]\Big| P^{X(t)} = P_t, 0 \le t \le 1, \\ X \in AC([0,1];\mathbb{R}^d), a.s.\Big\}.\end{aligned} \tag{3.100}$$

Then the following holds.

Theorem 3.7 (See [107]) *Suppose that (A.10) holds. Then for any* $\mathbf{P} := \{p(t,x)dx\}_{0\le t\le 1} \subset \mathscr{P}(\mathbb{R}^d)$ *for which* $\mathbf{v}_0(\mathbf{P})$ *is finite and* $D_x\sqrt{p(t,x)} \in L^2([0,1]\times\mathbb{R}^d, dtdx)$,

$$\mathbf{V}_\varepsilon(\mathbf{P}) = \mathbf{v}_\varepsilon(\mathbf{P}) \to \mathbf{V}_0(\mathbf{P}) = \mathbf{v}_0(\mathbf{P}), \quad \varepsilon \to 0. \tag{3.101}$$

In particular, any weak limit point of a minimizer of $\mathbf{V}_\varepsilon(\mathbf{P})$ *is a minimizer of* $\mathbf{V}_0(\mathbf{P})$.

Theorem 3.8 (See [107]) *Suppose that (A.10) holds and that L is strictly convex. Then for any* $\mathbf{P} := \{p(t,x)dx\}_{0\le t\le 1} \subset \mathcal{P}(\mathbb{R}^d)$ *for which* $\mathbf{v}_0(\mathbf{P})$ *is finite and* $D_x\sqrt{p(t,x)} \in L^2([0,1]\times\mathbb{R}^d, dtdx)$, $\mathbf{v}_0(\mathbf{P})$ *has the unique minimizer* $b_o(t,x)$ *and for any minimizer* $\{X(t)\}_{0\le t\le 1}$ *of* $\mathbf{V}_0(\mathbf{P})$,

$$X(t) = X(0) + \int_0^t b_o(s, X(s))ds, \quad t \in [0,1], \quad \text{a.s.}. \tag{3.102}$$

Proposition 3.7 (See [107]) *Suppose that* $L = |u|^2$. *Then for any* $\mathbf{P} := \{p(t,x)dx\}_{0\le t\le 1} \subset \mathcal{P}(\mathbb{R}^d)$ *such that* $\mathbf{v}_0(\mathbf{P})$ *is finite,* $D_x\sqrt{p(t,x)} \in L^2([0,1]\times\mathbb{R}^d, dtdx)$ *and for any* $M > 0$,

$$ess.inf\{p(t,x) : t \in [0,1], |x| \le M\} > 0, \tag{3.103}$$

the unique minimizer of $\mathbf{v}_0(\mathbf{P})$ *can be written as* $D_x u(t,x)$, *where* $u(t,\cdot) \in H^1_{loc}(\mathbb{R}^d)$ *dt–a.e..*

We explain that $\mathbf{V}_0(\mathbf{P})$ has a unique minimizer which is not random in the case where $d = 1$.

$$F_t(x) := \int_{(-\infty,x]} p(t,y)dy, \quad t \in [0,1], x \in \mathbb{R}, \tag{3.104}$$

$$F_t^{-1}(u) := \inf\{y \in \mathbb{R} : F_t(y) \ge u\}, \quad t \in [0,1], 0 < u < 1 \tag{3.105}$$

(see, e.g. [124]). Then the following holds.

Corollary 3.9 (See [107]) *Suppose that L is strictly convex and that (A.10) holds. If* $(t,x) \mapsto F_t(x)$ *is differentiable with locally bounded partial derivatives, then* $\mathbf{V}_0(\mathbf{P})$ *has the unique minimizer* $\lim_{s\in\mathbf{Q}\cap[0,1], s\to t} F_s^{-1}(F_0(X(0)))$.

In the case where a solution of the first-order ODE is not unique, it is not necessarily nonrandom. Indeed, the following is known.

Remark 3.8 (Salisbury's Problem) For a solution $X(t)$ of the first-order ODE, the problem of the nonrandomness of $X(t)$ (more precisely, the time invariance of $\sigma[X(s), 0 \le s \le t]$) is called Salisbury's problem. A counterexample was given by T. S. Salisbury (see [137]). In the case where $d \ge 2$, we do not know if a solution to (3.102) is random.

In our approach, we show the existence of a minimizer of a variational problem, i.e., we find a solution to the ODE with given marginals and also obtain optimal dynamics at the same time.

The superposition principle for (3.99) was generalized by L. Ambrosio (see also [5] and Theorem 2.3 in Sect. 2.2).

Theorem 3.9 (Ambrosio's Superposition Principle (See [3])) *Suppose that there exists $b : [0, 1] \times \mathbb{R}^d \longrightarrow \mathbb{R}^d$ and $\{P_t(dx)\}_{0 \le t \le 1} \subset \mathcal{P}(\mathbb{R}^d)$ such that (3.99) and the following hold:*

$$\int_0^1 dt \int_{\mathbb{R}^d} \frac{|b(t,x)|}{1+|x|} P_t(dx) < \infty. \tag{3.106}$$

Then there exists a stochastic process $\{X(t)\}_{0 \le t \le 1}$ for which the following holds:

$$X(t) = X(0) + \int_0^t b(s, X(s))ds, \tag{3.107}$$

$$P^{X(t)} = P_t, \quad 0 \le t \le 1. \tag{3.108}$$

3.2.3 SOT with a Nonconvex Cost

In this section, in the case where $d = 1$ and where a is not fixed, we consider slightly relaxed versions of the SOTs for which cost functions are not supposed to be convex. In this case, we need a generalization of D. Trevisan's result by V. I. Bogachev, M. Röckner, and S. V. Shaposhnikov [19]. As a fundamental problem of the stochastic optimal control theory, the test of the Markov property of a minimizer is known. We also discuss this problem for a finite horizon stochastic optimal control problem (see [61]).

Since a is not fixed in this section, we consider a new class of semimartingales. Let $u = \{u(t)\}_{0 \le t \le 1}$ and $\{W(t)\}_{0 \le t \le 1}$ be a progressively measurable real-valued process and a one-dimensional Brownian motion on a complete filtered probability space, respectively. In this section, the probability space under consideration is not fixed. Let $\sigma : [0, 1] \times \mathbb{R} \longrightarrow \mathbb{R}$ be a Borel measurable function. Let $Y^{u,\sigma} = \{Y^{u,\sigma}(t)\}_{0 \le t \le 1}$ be a continuous semimartingale such that the following holds weakly:

$$Y^{u,\sigma}(t) = Y^{u,\sigma}(0) + \int_0^t u(s)ds + \int_0^t \sigma(s, Y^{u,\sigma}(s))dW(s), \quad 0 \le t \le 1, \tag{3.109}$$

provided it exists.

Then the following is known.

Theorem 3.10 (See [105]) *Suppose that* $\Psi, L_1, L_2 \in C_b([0,1]\times\mathbb{R};[0,\infty))$. *Then for any* $r>0$ *and* $P_0 \in \mathscr{P}(\mathbb{R})$,

$$\inf_{\substack{(u,\sigma),\\ |\sigma|\ge r, P^{Y^{u,\sigma}(0)}=P_0}} E\left[\int_0^1 (L_1(t, Y^{u,\sigma}(t)) + L_2(t,u(t)))dt + \Psi(Y^{u,\sigma}(1))\right] \tag{3.110}$$

$$= \inf_{\substack{(u,\sigma),|\sigma|\ge r, P^{Y^{u,\sigma}(0)}=P_0,\\ u(t)=b_{Y^{u,\sigma}}(t,Y^{u,\sigma}(t))}} E\left[\int_0^1 (L_1(t, Y^{u,\sigma}(t)) + L_2(t,u(t)))dt + \Psi(Y^{u,\sigma}(1))\right],$$

where $b_{Y^{u,\sigma}}(t, Y^{u,\sigma}(t)) := E[u(t)|(t, Y^{u,\sigma}(t))]$.

For $r>0$,

$$\mathscr{U}_r := \left\{(u,\sigma)\middle| E\left[\int_0^1 \left(\frac{\sigma(t,Y^{u,\sigma}(t))^2}{1+|Y^{u,\sigma}(t)|^2} + |u(t)|\right)dt\right] < \infty, |\sigma|\ge r\right\}, \tag{3.111}$$

$$\mathscr{U}_{r,Mar} := \left\{(u,\sigma)\in\mathscr{U}_r | u(t) = b_{Y^{u,\rho}}(t, Y^{u,\sigma}(t)), 0\le t\le 1\right\}. \tag{3.112}$$

For $(u,\sigma)\in\mathscr{U}_r$,

$$F_t^{Y^{u,\sigma}}(x) := P(Y^{u,\sigma}(t)\le x), \tag{3.113}$$

$$G_t^u(x) := P(u(t)\le x), \tag{3.114}$$

$$\tilde{b}_{u,Y^{u,\sigma}}(t,x) := (G_t^u)^{-1}(1-F_t^{Y^{u,\sigma}}(x)), \quad (t,x)\in[0,1]\times\mathbb{R} \tag{3.115}$$

(see (3.105) for notation).

Remark 3.9 For $p\ge 1$, $P^{Y^{u,\sigma}(t)}(dx)\delta_{\tilde{b}_{u,Y^{u,\sigma}}(t,x)}(dy)$ is a (unique if $p>1$) maximizer of the following:

$$\sup\left\{\int_{\mathbb{R}^2} |x-y|^p \mu(dxdy)\middle|\mu\in\mathscr{A}(P^{Y^{u,\sigma}(t)}, P^{u(t)})\right\},$$

provided $x\mapsto F_t^{Y^{u,\sigma}}(x)$ is continuous and it is finite (see (2.1) for notation and [124, 142]).

For $(u,\sigma)\in\mathscr{U}_r$,

$$p^{Y^{u,\sigma}}(t,x) := \frac{P^{Y^{u,\sigma}(t)}(dx)}{dx} \tag{3.116}$$

exists dt–a.e. since r is positive and $(\sigma^2, b_{Y^{u,\sigma}}) \in \mathbf{A}_0(\{P^{Y^{u,\sigma}(t)}\}_{0\le t\le 1})$ from Remark 2.5 (see [18, p. 1042, Corollary 2.2.2]). Indeed, by Jensen's inequality,

$$\int_{\mathbb{R}} |b_{Y^{u,\sigma}}(t,y)| p^{Y^{u,\sigma}}(t,y)dy = E[|E[u(t)|(t, Y^{u,\sigma}(t))]|] \le E[|u(t)|]. \tag{3.117}$$

From the idea of covariance kernels (see [23, 24, 104, 109]),

$$\begin{aligned} &\tilde{a}_{u,Y^{u,\sigma}}(t,x) \\ &:= 1_{(0,\infty)}(p^{Y^{u,\sigma}}(t,x)) \frac{2\int_{-\infty}^{x} (\tilde{b}_{u,Y^{u,\sigma}}(t,y) - b_{Y^{u,\sigma}}(t,y)) p^{Y^{u,\sigma}}(t,y)dy}{p^{Y^{u,\sigma}}(t,x)}. \end{aligned} \tag{3.118}$$

The following will be proved later.

Theorem 3.11 (See [118]) *Let $r > 0$. For $(u,\sigma) \in \mathscr{U}_r$, there exists $\tilde{u}$ such that $(\tilde{u}, \tilde{\sigma} := (\sigma^2 + \tilde{a}_{u,Y^{u,\sigma}})^{1/2}) \in \mathscr{U}_{r,Mar}$ and that the following holds:*

$$P^{Y^{\tilde{u},\tilde{\sigma}}(t)} = P^{Y^{u,\sigma}(t)}, \quad t \in [0,1], \tag{3.119}$$

$$b_{Y^{\tilde{u},\tilde{\sigma}}} = \tilde{b}_{u,Y^{u,\sigma}}, \tag{3.120}$$

$$P^{b_{Y^{\tilde{u},\tilde{\sigma}}}(t, Y^{\tilde{u},\tilde{\sigma}}(t))} = P^{u(t)}, \quad dt\text{–a.e.}. \tag{3.121}$$

The following plays a crucial role in the proof of Theorem 3.11.

Theorem 3.12 (Superposition Principle (See [19])) *Let $\{P_t\}_{0\le t\le 1} \subset \mathscr{P}(\mathbb{R}^d)$. Suppose that there exists $(a,b) \in \mathbf{A}_0(\{P_t\}_{0\le t\le 1})$ such that the following holds:*

$$\int_0^1 dt \int_{\mathbb{R}^d} \frac{|a(t,x)| + |\langle x, b(t,x)\rangle|}{1+|x|^2} P_t(dx) < \infty. \tag{3.122}$$

Then Nelson's problem (3.88)–(3.90) has a solution.

For $r > 0$ and $\{P_t\}_{0\le t\le 1} \subset \mathscr{P}(\mathbb{R})$,

$$\begin{aligned} \mathbf{A}_{0,r}(\{P_t\}_{0\le t\le 1}) := \Bigg\{ (a,b) \in \mathbf{A}_0(\{P_t\}_{0\le t\le 1}) \Bigg| a \ge r^2, \\ \int_0^1 dt \int_{\mathbb{R}^d} \left(\frac{a(t,x)}{1+|x|^2} + |b(t,x)| \right) P_t(dx) < \infty \Bigg\}. \end{aligned} \tag{3.123}$$

Let $L_1, L_2 : [0,1] \times \mathbb{R} \longrightarrow [0,\infty)$ be Borel measurable. For (u,σ),

$$J(u,\sigma) := E\left[\int_0^1 (L_1(t, Y^{u,\sigma}(t)) + L_2(t, u(t)))dt \right]. \tag{3.124}$$

For $(a, b) \in \mathbf{A}_0(\{P_t\}_{0\le t\le 1})$,

$$I(\{P_t\}_{0\le t\le 1}, a, b) := \int_0^1 dt \int_{\mathbb{R}^d} (L_1(t, x) + L_2(t, b(t, x)))P_t(dx). \tag{3.125}$$

One easily obtains the following from Theorems 3.11 and 3.12.

Corollary 3.10 (See [118]) *Suppose that $L_1, L_2 : [0, 1] \times \mathbb{R} \longrightarrow [0, \infty)$ are Borel measurable. Then for any $r > 0$, the following holds. (i) For any $P_0, P_1 \in \mathscr{P}(\mathbb{R})$,*

$$\begin{aligned} &\inf\{J(u, \sigma)|(u, \sigma) \in \mathscr{U}_r, P^{Y^{u,\sigma}(t)} = P_t, t = 0, 1\} \qquad (3.126)\\ &= \inf\{J(u, \sigma)|(u, \sigma) \in \mathscr{U}_{r,Mar}, P^{Y^{u,\sigma}(t)} = P_t, t = 0, 1\}\\ &= \inf\{I(\{Q_t\}_{0\le t\le 1}, a, b)|(a, b) \in \mathbf{A}_{0,r}(\{Q_t\}_{0\le t\le 1}), Q_t = P_t, t = 0, 1\}. \end{aligned}$$

In particular, if there exists a minimizer in (3.126), then there exists a minimizer $(u, \sigma) \in \mathscr{U}_{r,Mar}$. (ii) For any $\{P_t\}_{0\le t\le 1} \subset \mathscr{P}(\mathbb{R})$,

$$\begin{aligned} &\inf\{J(u, \sigma)|(u, \sigma) \in \mathscr{U}_r, P^{Y^{u,\sigma}(t)} = P_t, 0 \le t \le 1\} \qquad (3.127)\\ &= \inf\{J(u, \sigma)|(u, \sigma) \in \mathscr{U}_{r,Mar}, P^{Y^{u,\sigma}(t)} = P_t, 0 \le t \le 1\}\\ &= \inf\{I(\{P_t\}_{0\le t\le 1}, a, b)|(a, b) \in \mathbf{A}_{0,r}(\{P_t\}_{0\le t\le 1})\}. \end{aligned}$$

In particular, if there exists a minimizer in (3.127), then there exists a minimizer $(u, \sigma) \in \mathscr{U}_{r,Mar}$.

Suppose that $L : [0, 1] \times \mathbb{R} \times \mathbb{R} \longrightarrow [0, \infty)$, $\Psi : \mathbb{R} \longrightarrow [0, \infty)$ are Borel measurable. Then for any $P_0 \in \mathscr{P}(\mathbb{R})$,

$$\begin{aligned} &\inf_{\substack{(u,\sigma)\in\mathscr{U}_r,\\ P^{Y^{u,\sigma}(0)}=P_0}} E\left[\int_0^1 L(t, Y^{u,\sigma}(t); u(t))dt + \Psi(Y^{u,\sigma}(1))\right] \qquad (3.128)\\ &= \inf_{P\in\mathscr{P}(\mathbb{R})} \left\{V_r(P_0, P) + \int_{\mathbb{R}} \Psi(x)P(dx)\right\}, \end{aligned}$$

where V_r denotes V with $\mathscr{A}$ replaced by $\{Y^{u,\sigma}|(u, \sigma) \in \mathscr{U}_r\}$. In particular, we easily obtain the following from Corollary 3.10.

Corollary 3.11 (See [118]) *In addition to the assumption of Corollary 3.10, suppose that $\Psi : \mathbb{R} \longrightarrow [0, \infty)$ is Borel measurable. Then for any $r > 0$ and $P_0 \in \mathscr{P}(\mathbb{R})$,*

$$\begin{aligned} &\inf\{J(u, \sigma) + E[\Psi(Y^{u,\sigma}(1))]|(u, \sigma) \in \mathscr{U}_r, P^{Y^{u,\sigma}(0)} = P_0\} \qquad (3.129)\\ &= \inf\{J(u, \sigma) + E[\Psi(Y^{u,\sigma}(1))]|(u, \sigma) \in \mathscr{U}_{r,Mar}, P^{Y^{u,\sigma}(0)} = P_0\}. \end{aligned}$$

In particular, if there exists a minimizer in (3.129), then there exists a minimizer $(u, \sigma) \in \mathscr{U}_{r,Mar}$.

We prove Theorem 3.11 by Theorem 3.12.

Proof of Theorem 3.11 For $(u, \sigma) \in \mathscr{U}_r$, the following holds (see [105]):

$$\tilde{a}_{u,Y^{u,\sigma}}(t, \cdot) \geq 0, \quad P^{\tilde{b}_{u,Y^{u,\sigma}}(t, Y^{u,\sigma}(t))} = P^{u(t)}, \quad dt\text{–a.e.}. \tag{3.130}$$

Indeed,

$$\int_{-\infty}^{x} \tilde{b}_{u,Y^{u,\sigma}}(t, y) p^{Y^{u,\sigma}}(t, y) dy = E[(G_t^u)^{-1}(1 - F_t^{Y^{u,\sigma}}(Y^{u,\sigma}(t))); Y^{u,\sigma}(t) \leq x],$$

$$\begin{aligned}\int_{-\infty}^{x} b_{Y^{u,\sigma}}(t, y) p^{Y^{u,\sigma}}(t, y) dy &= E[E[u(t)|(t, Y^{u,\sigma}(t))]; Y^{u,\sigma}(t) \leq x] \\ &= E[u(t); Y^{u,\sigma}(t) \leq x].\end{aligned}$$

For an $\mathbb{R}^2$-valued random variable $Z = (X, Y)$ on a probability space,

$$\begin{aligned}E[Y; X \leq x] &= \int_0^{\infty} \{F_X(x) - F_Z(x, y)\} dy - \int_{-\infty}^{0} F_Z(x, y) dy, \\ F_Z(x, y) &\geq \max(F_X(x) + F_Y(y) - 1, 0) \\ &= P(F_X^{-1}(U) \leq x, F_Y^{-1}(1 - U) \leq y),\end{aligned}$$

where F_X denotes the distribution function of X and U is a uniformly distributed random variable on $[0, 1]$. The distribution functions of $F_X^{-1}(U)$ and $F_Y^{-1}(1 - U)$ are F_X and F_Y, respectively. From (3.116), $F_t^{Y^{u,\sigma}}(Y^{u,\sigma}(t))$ is uniformly distributed on $[0, 1]$ and $(F_t^{Y^{u,\sigma}})^{-1}(F_t^{Y^{u,\sigma}}(Y^{u,\sigma}(t))) = Y^{u,\sigma}(t)$, P–a.s., dt–a.e. (see [46] or, e.g. [124, 130]).

It is easy to see that the following holds from (3.117) and (3.130):

$$(\sigma^2 + \tilde{a}_{u,Y^{u,\sigma}}, \tilde{b}_{u,Y^{u,\sigma}}) \in \mathbf{A}_0(\{P^{Y^{u,\sigma}(t)}\}_{0 \leq t \leq 1}).$$

Indeed, from (3.130), the following holds:

$$E\left[\int_0^1 |\tilde{b}_{u,Y^{u,\sigma}}(t, Y^{u,\sigma}(t))| dt\right] = E\left[\int_0^1 |u(t)| dt\right] < \infty. \tag{3.131}$$

The following will be proved below:

$$E\left[\int_0^1 \frac{\tilde{a}_{u,Y^{u,\sigma}}(t, Y^{u,\sigma}(t))}{1 + |Y^{u,\sigma}(t)|^2} dt\right] < \infty. \tag{3.132}$$

Equations (3.131)–(3.132) and Theorem 3.12 complete the proof. We prove (3.132).

$$E\left[\int_0^1 \frac{\tilde{a}_{u,Y^{u,\sigma}}(t, Y^{u,\sigma}(t))}{2(1+|Y^{u,\sigma}(t)|^2)}dt\right] \tag{3.133}$$

$$= \int_0^1 dt \int_{\mathbb{R}} \frac{1}{1+x^2}dx \int_{-\infty}^{x} (\tilde{b}_{u,Y^{u,\sigma}}(t,y) - b_{Y^{u,\sigma}}(t,y))p^{Y^{u,\sigma}}(t,y)dy.$$

From (3.130),

$$\int_{-\infty}^{\infty} \tilde{b}_{u,Y^{u,\sigma}}(t,y)p^{Y^{u,\sigma}}(t,y)dy = E[u(t)] = E[E[u(t)|(t, Y^{u,\sigma}(t))]] \tag{3.134}$$

$$= \int_{-\infty}^{\infty} b_{Y^{u,\sigma}}(t,y)p^{Y^{u,\sigma}}(t,y)dy, \quad dt\text{–a.e..}$$

In particular, the following holds dt–a.e.:

$$\int_{\mathbb{R}} \frac{1}{1+x^2}dx \int_{-\infty}^{x} (\tilde{b}_{u,Y^{u,\sigma}}(t,y) - b_{Y^{u,\sigma}}(t,y))p^{Y^{u,\sigma}}(t,y)dy \tag{3.135}$$

$$= \int_{-\infty}^{0} \frac{1}{1+x^2}dx \int_{-\infty}^{x} (\tilde{b}_{u,Y^{u,\sigma}}(t,y) - b_{Y^{u,\sigma}}(t,y))p^{Y^{u,\sigma}}(t,y)dy$$

$$- \int_{0}^{\infty} \frac{1}{1+x^2}dx \int_{x}^{\infty} (\tilde{b}_{u,Y^{u,\sigma}}(t,y) - b_{Y^{u,\sigma}}(t,y))p^{Y^{u,\sigma}}(t,y)dy$$

$$= \int_{-\infty}^{0} (\tilde{b}_{u,Y^{u,\sigma}}(t,y) - b_{Y^{u,\sigma}}(t,y))p^{Y^{u,\sigma}}(t,y)dy \int_{y}^{0} \frac{1}{1+x^2}dx$$

$$- \int_{0}^{\infty} (\tilde{b}_{u,Y^{u,\sigma}}(t,y) - b_{Y^{u,\sigma}}(t,y))p^{Y^{u,\sigma}}(t,y)dy \int_{0}^{y} \frac{1}{1+x^2}dx$$

$$\leq \int_{\mathbb{R}} |\arctan y|(|\tilde{b}_{u,Y^{u,\sigma}}(t,y)| + |b_{Y^{u,\sigma}}(t,y)|)p^{Y^{u,\sigma}}(t,y)dy.$$

Since $|\arctan y|$ is bounded, (3.117) and (3.131) completes the proof of (3.132).

References

1. Adams, S., Dirr, N., Peletier, M.A., Zimmer, J.: From a large-deviations principle to the Wasserstein gradient flow: a new micro-macro passage. Commun. Math. Phys. **307**(3), 791–815 (2011)
2. Albeverio, S., Yasue, K., Zambrini, J.C.: Euclidean quantum mechanics: analytical approach. Ann. Inst. H. Poincaré Phys. Théor. **50**(3), 259–308 (1989)
3. Ambrosio, L.: Transport equation and Cauchy problem for BV vector fields. Invent. Math. **158**(2), 227–260 (2004)
4. Ambrosio, L., Pratelli, A: Existence and stability results in the L^1 theory of optimal transportation. In: Caffarelli, L.A., Salsa, S. (eds.) Optimal Transportation and Applications (CIME Series, Martina Franca, 2001), Lecture Notes in Math., vol. 1813, pp. 123–160. Springer, Heidelberg (2003)
5. Ambrosio, L., Trevisan, D.: Well-posedness of Lagrangian flows and continuity equations in metric measure spaces. Anal. PDE **7**(5), 1179–1234 (2014)
6. Aronson, D.G.: Bounds for the fundamental solution of a parabolic equation. Bull. Am. Math. Soc. **73**, 890–896 (1967)
7. Backhoff, J., Conforti, G., Gentil, I., Léonard, C.: The mean field Schrödinger problem: ergodic behavior, entropy estimates and functional inequalities. Probab. Theory Related Fields (2020). https://doi.org/10.1007/s00440-020-00977-8
8. Bakelman, I.J.: Convex Analysis and Nonlinear Geometric Elliptic Equations. Springer, Heidelberg (1994)
9. Benamou, J.D., Brenier, Y.: A numerical method for the optimal mass transport problem and related problems. In: Caffarelli, L.A., Milman, M. (eds.) Monge Ampére Equation: Applications to Geometry and Optimization, Proceedings of the NSF–CBMS Conference, Deerfield Beach, FL, 1997, Contemporary Mathematics, vol. 226, pp. 1–11. American Mathematical Society, Providence, RI (1999)
10. Bensoussan, A., Frehse, J., Yam, P.: Mean Field Games and Mean Field Type Control Theory. Springer Briefs in Mathematics. Springer, Heidelberg (2013)
11. Bensoussan, A., Frehse, J., Yam, P.: On the interpretation of the master equation. Stoch. Process. Appl. **127**(7), 2093–2137 (2017)
12. Bernstein, S.: Sur les liaisons entre les grandeurs alétoires. Verh. des intern. Mathematikerkongr. Zurich 1932, Band **1**, 288–309 (1932)
13. Beurling, A.: An automorphism of product measures. Ann. Math. **72**, 189–200 (1960)
14. Billingsley, P.: Convergence of Probability Measures. Wiley–Interscience, New York (1999)
15. Blumenthal, R.M., Corson, H.H.: On continuous collections of measures. In: Le Cam, L. et al. (eds.) Proc. 6th Berkeley Sympos. Math. Statist. Probab. 2, Berkeley 1970/1971, pp. 33–40. University of California Press, Berkeley (1972)

T. Mikami, *Stochastic Optimal Transportation*, SpringerBriefs in Mathematics,
https://doi.org/10.1007/978-981-16-1754-6

16. Bogachev, V.I., Kolesnikov, A.V., Medvedev, K.V.: On triangular transformations of measures (Russian). Dokl. Akad. Nauk **396**(6), 727–732 (2004)
17. Bogachev, V.I., Kolesnikov, A.V., Medvedev, K.V.: Triangular transformations of measures (Russian). Sb. Math. **196**(3–4), 309–335 (2005)
18. Bogachev, V.I., Krylov, N.V., Röckner, M.: Elliptic and parabolic equations for measures. Russian Math. Surv. **64**(6), 973–1078 (2009)
19. Bogachev, V.I., Röckner, M., Shaposhnikov, S.V.: On the Ambrosio–Figalli–Trevisan superposition principle for probability solutions to Fokker–Planck–Kolmogorov equations. J. Dynam. Differential Equations (2020). https://doi.org/10.1007/s10884-020-09828-5
20. Bowles, M., Ghoussoub, N.: A theory of transfers: duality and convolution (2018). arXiv:1804.08563
21. Brenier, Y.: Décomposition polaire et réarrangement monotone des champs de vecteurs. C. R. Acad. Sci. Paris Série I **305**(19), 805–808 (1987)
22. Brenier, Y.: Polar factorization and monotone rearrangement of vector-valued functions. Commun. Pure Appl. Math. **44**(4), 375–417 (1991)
23. Cacoullos, T., Papathanasiou, V., Utev, S.A.: Another characterization of the normal distribution and a related proof of the central limit theorem. Theory Probab. Appl. **37**(4), 581–588 (1992)
24. Cacoullos, T., Papathanasiou, V., Utev, S.A.: Variational inequalities with examples and an application to the central limit theorem. Ann. Probab. **22**(3), 1607–1618 (1994)
25. Caffarelli, L.A., Salsa, S. (eds.): Optimal Transportation and Applications (CIME Series, Martina Franca, 2001), Lecture Notes in Math., vol. 1813. Springer, Heidelberg (2003)
26. Caffarelli, L.A., Feldman, M., McCann, R.J.: Constructing optimal maps for Monge's transport problem as a limit of strictly convex costs. J. Am. Math. Soc. **15**(1), 1–26 (2001)
27. Cardaliaguet, P.: Notes on Mean Field Games (from P.- L. Lions' lectures at College de France). January 15, 2012
28. Carlen, E.A.: Conservative diffusions. Commun. Math. Phys. **94**(3), 293–315 (1984)
29. Carlen, E.A.: Existence and sample path properties of the diffusions in Nelson's stochastic mechanics. In: Albeverio, S., Blanchard, Ph., Streit, L. (eds.) Stochastic Processes–Mathematics and Physics, Bielefeld 1984, Lecture Notes in Math., vol. 1158, pp. 25–51. Springer, Heidelberg (1986)
30. Carlier, G., Galichon, F., Santambrogio, F.: From Knothe's transport to Brenier's map and a continuation method for optimal transport. SIAM J. Math. Anal. **41**(6), 2554–2576 (2010)
31. Carmona, R.: Probabilistic construction of Nelson processes. In: Itô, K., Ikeda, N. (eds.) Proc. Probabilistic Methods in Mathematical Physics, Katata 1985, pp. 55–81. Kinokuniya, Tokyo (1987)
32. Cattiaux, P., Léonard, C.: Minimization of the Kullback information of diffusion processes. Ann. Inst. H. Poincaré Probab. Stat. **30**(1), 83–132 (1994)
33. Cattiaux, P., Léonard, C.: Correction to: "Minimization of the Kullback information of diffusion processes" [Ann. Inst. H. Poincaré Probab. Stat. **30**(1), 83–132 (1994)]. Ann. Inst. H. Poincaré Probab. Stat. **31**(4), 705–707 (1995)
34. Cattiaux, P., Léonard, C.: Large deviations and Nelson processes. Forum Math. **7**(1), 95–115 (1995)
35. Cattiaux, P., Léonard, C.: Minimization of the Kullback information for some Markov processes. In: Azema, J., et al. (eds.) Séminaire de Probabilités, XXX, Lecture Notes in Math., vol. 1626, pp. 288–311. Springer, Heidelberg (1996)
36. Chen, Y., Georgiou, T., Pavon, M.: Entropic and displacement interpolation: a computational approach using the Hilbert metric. SIAM J. Appl. Math. **76**(6), 2375–2396 (2016)
37. Chen, Y., Georgiou, T.T., Pavon, M.: Stochastic control liaisons: Richard Sinkhorn meets Gaspard Monge on a Schrödinger bridge (2020). arXiv: 2005.10963
38. Chung, K.L., Zambrini, J.C.: Introduction to Random Time and Quantum Randomness. World Scientific Publishing Co. Inc., Singapore (2003)

39. Conforti, G.: A second order equation for Schrödinger bridges with applications to the hot gas experiment and entropic transportation cost. Probab. Theory Related Fields **174**(1–2), 1–47 (2019)
40. Conforti, G., Ripani, L.: Around the entropic Talagrand inequality. Bernoulli **26**(2), 1431–1452 (2020)
41. Cover, T.M., Thomas, J.A.: Elements of Information Theory, 2nd edn. Wiley-Interscience, New York (2006)
42. Crandall, M.G., Ishii, H., Lions, P.-L.: User's guide to viscosity solutions of second order partial differential equations. Bull. Amer. Math. Soc. **27**(1), 1–67 (1992)
43. Cruzeiros, A.B., Zambrini, J.C.: Euclidean quantum mechanics. An outline. In: Cardoso, A.I., de Faria, M., Potthoff, J., Sénéor, R., Streit, L. (eds.) Stochastic Analysis and Applications in Physics. NATO ASI Series (Series C: Mathematical and Physical Sciences), vol. 449, pp. 59–97. Springer, Dordrecht (1994)
44. Cuturi, M.: Sinkhorn distances: Lightspeed computation of optimal transportation. In: Burges, C.J.C., Bottou, L., Welling, M., Ghahramani, Z., Weinberger, K.Q. (eds.) Neural Information Processing Systems 2013. Advances in Neural Information Processing Systems, vol. 26, pp. 2292–2300 (2013)
45. Dai Pra, P.: A stochastic control approach to reciprocal diffusion processes. Appl. Math. Optim. **23**(3), 313–329 (1991)
46. Dall'Aglio, G.: Sugli estremi dei momenti delle funzioni di ripartizione doppia. Ann. Scuola Norm. Sup. Pisa Cl. Sci. (3) **10**, 35–74 (1956)
47. Dall'Aglio, G., Kotz, S., Salinetti, G. (eds.): Advances in Probability Distributions with Given Marginals. Mathematics and its Applications, vol. 67. Kluwer Academic Publishers, Boston (1991)
48. Deuschel, J.D., Stroock, D.W.: Large Deviations. Academic Press, Inc., Boston (1989)
49. Doob, J.L.: Conditional Brownian motion and the boundary limits of harmonic functions. Bull. Soc. Math. France **85**, 431–458 (1957)
50. Dudley, R.M.: Probabilities and Metrics. Convergence of Laws on Metric Spaces, with a View to Statistical Testing, Lecture Notes Series, vol. 45. Matematisk Institut, Aarhus Universitet, Aarhus (1976)
51. Duong, M.H., Laschos, V., Renger, M.: Wasserstein gradient flows from large deviations of many-particle limits. ESAIM Control Optim. Calc. Var. **19**(4), 1166–1188 (2013). Erratum at www.wias-berlin.de/people/renger/Erratum/DLR2015ErratumFinal.pdf
52. Dupuis, P., Ellis, R.S.: A Weak Convergence Approach to the Theory of Large Deviations. John Wiley & Sons, New York (1997)
53. Erbar, M., Maas, J., Renger, M.: From large deviations to Wasserstein gradient flows in multiple dimensions. Electron. Commun. Probab. **20**(89), 12 pp. (2015)
54. Evans, L.C.: Partial differential equations and Monge–Kantorovich mass transfer. In: Current Developments in Mathematics, 1997 (Cambridge, MA), pp. 65–126. International Press, Boston (1997)
55. Evans, L.C.: Partial Differential Equations. AMS, Providence (1998)
56. Evans, L.C., Gangbo, W.: Differential equations methods for the Monge–Kantorovich mass transfer problem. Mem. Amer. Math. Soc. **137**(653) (1999)
57. Feng, J., Nguyen, T.: Hamilton-Jacobi equations in space of measures associated with a system of conservation laws. J. Math. Pures Appl. (9) **97**(4), 318–390 (2012)
58. Feyel, D., Üstünel, A.S.: Monge–Kantorovitch measure transportation and Monge–Ampère equation on Wiener space. Probab. Theory Related Fields **128**(3), 347–385 (2004)
59. Figalli, A.: Existence and uniqueness of martingale solutions for SDEs with rough or degenerate coefficients. J. Funct. Anal. **254**(1), 109–153 (2008)
60. Fleming, W.H., Rishel, R.W.: Deterministic and Stochastic Optimal Control. Springer, New York (1975)
61. Fleming, W.H., Soner, H.M.: Controlled Markov Processes and Viscosity Solutions. Springer, New York (1993)

62. Föllmer, H.: Random fields and diffusion processes. In: Hennequin, P.L. (ed.) École d'Été de Probabilités de Saint-Flour XV–XVII, 1985–87, Lecture Notes in Math., vol. 1362, pp. 101–203. Springer, Heidelberg (1988)
63. Fortet, R.: Résolution d'un Système d'équations de M. Schrödinger, J. Math. Pures Appl. (9) **19**, 83–105 (1940)
64. Freidlin, M.I., Wentzell, A.D.: Random Perturbations of Dynamical Systems. Springer, New York (1984)
65. Friedman, A.: Stochastic Differential Equations and Applications. Dover Publications, New York (2006)
66. Friedman, A.: Partial Differential Equations of Parabolic Type. Dover Publications, New York (2013)
67. Gangbo, W., McCann, R.J.: The geometry of optimal transportation. Acta Math. **177**(2), 113–161 (1996)
68. Garling, D.J.H.: A Course in Mathematical Analysis: Volume 3, Complex Analysis, Measure and Integration. Cambridge University Press, Cambridge (2014)
69. Gentil, I., Léonard, C., Ripani, L., Tamanini, L.: An entropic interpolation proof of the HWI inequality. Stoch. Process. Appl. **130**(2), 907–923 (2020)
70. Gomes, D.A.: A stochastic analogue of Aubry–Mather theory. Nonlinearity **15**(3), 581–603 (2002)
71. Gomes, D.A., Mitake, H, Tran, H.V.: The large time profile for Hamilton–Jacobi–Bellman equations (2020). arXiv:2006.04785
72. Gutiérrez, C.E.: The Monge–Ampère Eqaution. Birkhäuser, Boston (2001)
73. Haussmann, U.G., Pardoux, E.: Time reversal of diffusions. Ann. Probab. **14**(4), 1188–1205 (1986)
74. Hernández-Lerma, O., Gabriel, J.R.: Strong duality of the Monge–Kantorovich mass transfer problem in metric spaces. Math. Z. **239**(3), 579–591 (2002)
75. Ikeda, N., Watanabe, S.: Stochastic Differential Equations and Diffusion Processes. North-Holland/Kodansha, Tokyo (1981)
76. Ioffe, A.D., Tihomirov, V.M.: Theory of Extremal Problems. North-Holland, Amsterdam (1979)
77. Ito, H.M., Mikami, T.: Poissonian asymptotics of a randomly perturbed dynamical system: Flip-flop of the stochastic disk dynamo. J. Statist. Phys. **85**(1–2), 41–53 (1996)
78. Jacod, J., Shiryaev, A.N.: Limit Theorems for Stochastic Processes. Springer, Heidelberg (1987)
79. Jamison, B.: Reciprocal processes. Z. Wahrscheinlichkeitstheorie und Verw. Gebiete **30**, 65–86 (1974)
80. Jamison, B.: The Markov process of Schrödinger. Z. Wahrscheinlichkeitstheorie und Verw. Gebiete **32**, 323–331 (1975)
81. Jordan, R., Kinderlehrer, D., Otto, F.: The variational formulation of the Fokker–Planck equation. SIAM J. Math. Anal. **29**(1), 1–17 (1998)
82. Kamae, T., Krengel, U.: Stochastic partial ordering. Ann. Probab. **6**(6), 1044–1049 (1978).
83. Kantorovich, L.V.: On mass transportation. C. R. (Doklady) Acad. Sci. URSS (N.S.) **37**(7–8), 199–201 (1942)
84. Kantorovich, L.V.: On a problem of Monge. C. R. (Doklady) Acad. Sci. URSS (N.S.) **3**(2), 225–226 (1948)
85. Kellerer, H.G.: Duality theorems for marginal problems. Z. Wahrsch. Verw. Gebiete **67**(4), 399–432 (1984)
86. Kellerer, H.G.: Representation of Markov kernels by random mappings under order conditions. In: Benes, V., Stepan, J. (eds.) Distributions with Given Marginals and Moment Problems, Proceedings, Prague 1996, pp. 143–160. Kluwer Academic Publishers, Boston (1997)
87. Knothe, H.: Contributions to the theory of convex bodies. Michigan Math. J. **4**, 39–52 (1957)
88. Knott, M., Smith, C.S.: On the optimal mapping of distributions. J. Optim. Theory Appl. **43**(1), 39–49 (1984)

89. Koike, S.: A beginner's guide to the theory of viscosity solutions. MSJ Memoirs, vol. 13. Mathematical Society Japan, Tokyo (2004)
90. Krener, A.J.: Reciprocal diffusions and stochastic differential equations of second order. Stochastics **24**(4), 393–422 (1988)
91. Krener, A.J., Frezza, R., Levy, B.C.: Gaussian reciprocal processes and selfadjoint stochastic differential equations of second order. Stoch. Stoch. Rep. **34**(1–2), 29–56 (1991)
92. Krylov, N.V.: Controlled Diffusion Processes. Springer, Heidelberg (1980)
93. Lasry, J.-M., Lions, P.-L.: Mean field games. Jpn. J. Math. **2**(1), 229–260 (2007)
94. Lassalle, R., Cruzeiro, A.B.: An intrinsic calculus of variations for functionals of laws of semi-martingales. Stoch. Process. Appl. **129**(10), 3585–3618 (2019)
95. Léonard, C.: From the Schrödinger problem to the Monge–Kantorovich problem. J. Funct. Anal. **262**(4), 1879–1920 (2012)
96. Léonard, C.: A survey of the Schrödinger problem and some of its connections with optimal transport. Special Issue on Optimal Transport and Applications. Discrete Contin. Dyn. Syst. **34**(4), 1533–1574 (2014)
97. Levin, V.L.: Duality and approximation in the problem of mass transfer. In: Mityagin, B.S. (ed.) Mathematical Economics and Functional Analysis, pp. 94–108. Izdat. "Nauka", Moscow (1974)
98. Levin, V.L.: General Monge–Kantorovich problem and its applications in measure theory and mathematical economics. In: Leifman, L.J. (ed.) Functional Analysis, Optimization, and Mathematical Economics, pp. 141–176. Oxford University Press, London (1990)
99. Lévy, P.: Théorie de l'Addition des Variables Aléatoires. pp. 71–73, 121–123. Gauthier–Villars, Paris (1937)
100. Liptser, R.S., Shiryaev, A.N.: Statistics of Random Processes I. Springer, Heidelberg (1977)
101. Liu, C., Neufeld, A.: Compactness criterion for semimartingale laws and semimartingale optimal transport. Trans. Am. Math. Soc. **372**(1), 187–231 (2019)
102. McCann, R.J.: A convexity principle for interacting gases. Adv. Math. **128**(1), 153–179 (1997)
103. Mikami, T.: Variational processes from the weak forward equation. Commun. Math. Phys. **135**(1), 19–40 (1990)
104. Mikami, T.: Equivalent conditions on the central limit theorem for a sequence of probability measures on $\mathbb{R}$. Statist. Probab. Lett. **37**(3), 237–242 (1998)
105. Mikami, T.: Markov marginal problems and their applications to Markov optimal control. In: McEneaney, W.M. et al. (eds.) Stochastic Analysis, Control, Optimization and Applications, A Volume in Honor of W. H. Fleming, pp. 457–476. Birkhäuser, Boston (1999)
106. Mikami, T.: Dynamical systems in the variational formulation of the Fokker–Planck equation by the Wasserstein metric. Appl. Math. Optim. **42**(2), 203–227 (2000)
107. Mikami, T.: Optimal control for absolutely continuous stochastic processes and the mass transportation problem. Electron. Commun. Probab. **7**, 199–213 (2002)
108. Mikami, T.: Monge's problem with a quadratic cost by the zero-noise limit of h-path processes. Probab. Theory Related Fields **129**(2), 245–260 (2004)
109. Mikami, T.: Covariance kernel and the central limit theorem in the total variation distance. J. Multivariate Anal. **90**(2), 257–268 (2004)
110. Mikami, T.: A simple proof of duality theorem for Monge–Kantorovich problem. Kodai Math. J. **29**(1), 1–4 (2006)
111. Mikami, T.: Semimartingales from the Fokker–Planck equation. Appl. Math. Optim. **53**(2), 209–219 (2006)
112. Mikami, T.: Marginal problem for semimartingales via duality. In: Giga, Y., Ishii, K., Koike, S., et al. (eds.) International Conference for the 25th Anniversary of Viscosity Solutions, pp. 133–152. Gakuto International Series. Mathematical Sciences and Applications, vol. 30. Gakkotosho, Tokyo (2008)
113. Mikami, T.: Optimal transportation problem as stochastic mechanics. In: Selected Papers on Probability and Statistics, Amer. Math. Soc. Transl. Ser. 2, vol. 227, pp. 75–94. American Mathematical Society, Providence, RI (2009)

114. Mikami, T.: A characterization of the Knothe–Rosenblatt processes by a convergence result. SIAM J. Control Optim. **50**(4), 1903–1920 (2012)
115. Mikami, T.: Two end points marginal problem by stochastic optimal transportation. SIAM J. Control Optim. **53**(4), 2449–2461 (2015)
116. Mikami, T.: Regularity of Schrödinger's functional equation and mean field PDEs for h-path processes. Osaka J. Math. **56**(4), 831–842 (2019)
117. Mikami, T.: Regularity of Schrödinger's functional equation in the weak topology and moment measures. J. Math. Soc. Japan. **73**(1), 99–123 (2021)
118. Mikami, T.: Stochastic optimal transport revisited. SN Partial Differ. Equ. Appl. **2**, 5 (2021)
119. Mikami, T., Thieullen, M.: Duality theorem for stochastic optimal control problem. Stoch. Process. Appl. **116**(12), 1815–1835 (2006)
120. Mikami, T., Thieullen, M.: Optimal transportation problem by stochastic optimal control. SIAM J. Control Optim. **47**(3), 1127–1139 (2008)
121. Monge, G.: Mémoire sur la théorie des déblais et des remblais. In: Histoire de l'Académie Royale des Sciences de Paris, pp. 666–704 (1781)
122. Nagasawa, M.: Time reversions of Markov processes. Nagoya Math. J. **24**, 177–204 (1964)
123. Nagasawa, M.: Transformations of diffusion and Schrödinger process. Probab. Theory Related Fields **82**(1), 109–136 (1989)
124. Nelsen, R.B.: An Introduction to Copulas, 2nd edn. Springer, Heidelberg (2006)
125. Nelson, E.: Dynamical Theories of Brownian Motion. Princeton University Press, Princeton (1967)
126. Nelson, E.: Quantum Fluctuations. Princeton University Press, Princeton (1984)
127. Nielsen, F, Barbaresco, F. (eds.): Geometric Science of Information: Second International Conference, GSI 2015, Palaiseau, France, October 28–30, 2015, Proceedings (Lecture Notes in Computer Science Book 9389). Springer, Heidelberg (2015)
128. Pal, S.: On the difference between entropic cost and the optimal transport cost (2019). arXiv:1905.12206v1
129. Peyre, G., Cuturi, M.: Computational Optimal Transport: With Applications to Data Science. Now Publishers, Boston (2019)
130. Rachev, S.T., Rüschendorf, L.: Mass Transportation Problems, Vol. I: Theory, Vol. II: Application. Springer, Heidelberg (1998)
131. Ripani, L.: Convexity and regularity properties for entropic interpolations. J. Funct. Anal. **277**(2), 368–391 (2019)
132. Röckner, M., Zhang, X.: Well-posedness of distribution dependent SDEs with singular drifts (2018). arXiv:1809.02216
133. Röckner, M., Xie, L., Zhang, X.: Superposition principle for non-local Fokker–Planck–Kolmogorov operators. Probab. Theory Related Fields **178**(3–4), 699–733 (2020)
134. Rosenblatt, M.: Remarks on a multivariate transformation. Ann. Math. Stat. **23**, 470–472 (1952)
135. Rüschendorf, L., Thomsen, W.: Note on the Schrödinger equation and I-projections. Stat. Probab. Lett. **17**(5), 369–375 (1993)
136. Sakurai, J.J., Napolitano, J.J.: Modern Quantum Mechanics. Addison Wesley, Boston (2010)
137. Salisbury, T.S.: An increasing diffusion. In: Cinlar, E., Chung, K.L., Getoor, R.K. (eds.) Seminar on Stochastic Processes 1984, pp. 173–194. Birkhäuser, Boston (1986)
138. Santambrogio, F.: Dealing with moment measures via entropy and optimal transport. J. Funct. Anal. **271**(2), 418–436 (2016)
139. Santambrogio, F.: Optimal Transport for Applied Mathematicians: Calculus of Variations, PDEs, and Modeling. Birkhäuser, Boston (2016)
140. Schrödinger, E.: Ueber die Umkehrung der Naturgesetze. Sitz. Ber. der Preuss. Akad. Wissen., Berlin, Phys. Math., pp. 144–153 (1931)
141. Schrödinger, E.: Théorie relativiste de l'electron et l'interprétation de la mécanique quantique. Ann. Inst. H. Poincaré **2**(4), 269–310 (1932)
142. Schweizer, B., Sklar, A.: Probabilistic Metric Space. Dover Publications, New York (2005)

143. Sheu, S.J.: Some estimates of the transition density of a nondegenerate diffusion Markov process. Ann. Probab. **19**(2), 538–561 (1991)
144. Smith, C.S., Knott, M.: Note on the optimal transportation of distributions. J. Optim. Theory Appl. **52**(2), 323–329 (1987)
145. Smith, C., Knott, M.: On Hoeffding–Fréchet bounds and cyclic monotone relations. J. Multivariate Anal. **40**(2), 328–334 (1992)
146. Talagrand, M.: Transportation cost for Gaussian and other product measures. Geom. Funct. Anal. **6**(3), 587–600 (1996)
147. Tan, X., Touzi, N.: Optimal transportation under controlled stochastic dynamics. Ann. Probab. **41**(5), 3201–3240 (2013)
148. Tanaka, H.: An inequality for a functional of probability distributions and its application to Kac's one-dimensional model of a Maxwellian gas. Z. Wahrscheinlichkeitstheorie und Verw. Gebiete **27**, 47–52 (1973)
149. Thieullen, M.: Second order stochastic differential equations and non-Gaussian reciprocal diffusions. Probab. Theory Related Fields **97**(1–2), 231–257 (1993)
150. Trevisan, D.: Well-posedness of multidimensional diffusion processes with weakly differentiable coefficients. Electron J. Probab. **21**(22), 41 pp. (2016)
151. Trudinger, N.S., Wang, X.-J.: On the Monge mass transfer problem. Calc. Var. Partial Differential Equations **13**(1), 19–31 (2001)
152. Villani, C.: Topics in Optimal Transportation. American Mathematical Society, Providence, RI (2003)
153. Villani, C.: Optimal Transport: Old and New. Springer, Heidelberg (2008)
154. von Renesse, M.K., Sturm, K.T.: Transport inequalities, gradient estimates, entropy, and Ricci curvature. Commun. Pure Appl. Math. **58**(7), 923–940 (2005)
155. Wakolbinger, A.: Schrödinger Bridges from 1931 to 1991. In: Cabaña, E., et al. (eds.) Proc. of the 4th Latin American Congress in Probability and Mathematical Statistics, Mexico City 1990, Contribuciones en probabilidad y estadistica matemática 3, pp. 61–79 (1992)
156. Zambrini, J.C.: Variational processes. In: Albeverio, S. et al. (eds.) Stochastic Processes in Classical and Quantum Systems, Ascona 1985, Lecture Notes in Physics, vol. 262, pp. 517–529. Springer, Heidelberg (1986)
157. Zambrini, J.C.: Variational processes and stochastic versions of mechanics. J. Math. Phys. **27**(9), 2307–2330 (1986)
158. Zambrini, J.C.: Stochastic mechanics according to E. Schrödinger. Phys. Rev. A (3) **33**(3), 1532–1548 (1986)
159. Zheng, W.A.: Tightness results for laws of diffusion processes application to stochastic mechanics. Ann. Inst. Henri Poincaré Probab. Stat. **21**(2), 103–124 (1985)

Printed in the United States
by Baker & Taylor Publisher Services